T0237705

Lecture Notes in Artificial Intelligence   10839

Subseries of Lecture Notes in Computer Science

More information about this series at http://www.springer.com/series/1244

Jaap van den Herik · Ana Paula Rocha
Joaquim Filipe (Eds.)

Agents and Artificial Intelligence

9th International Conference, ICAART 2017
Porto, Portugal, February 24–26, 2017
Revised Selected Papers

 Springer

Editors
Jaap van den Herik
Leiden University
Leiden
The Netherlands

Ana Paula Rocha
University of Porto
Porto
Portugal

Joaquim Filipe
INSTICC
Polytechnic Institute of Setúbal
Setubal
Portugal

ISSN 0302-9743   ISSN 1611-3349   (electronic)
Lecture Notes in Artificial Intelligence
ISBN 978-3-319-93580-5   ISBN 978-3-319-93581-2   (eBook)
https://doi.org/10.1007/978-3-319-93581-2

Library of Congress Control Number: 2018947333

LNCS Sublibrary: SL7 – Artificial Intelligence

Preface

The present book includes 11 extended and revised versions of a set of selected papers from the 9th International Conference on Agents and Artificial Intelligence (ICAART 2017), held in Porto, Portugal, during February 24–26, 2017.

ICAART 2017 received 158 paper submissions from 45 countries, of which 7% are included in this book.

The selection procedure consisted of two phases. In the first phase, the event chairs selected a set of potentially eligible papers. Their selection was based on a number of criteria that included the classifications and comments provided by the Program Committee members, the session chairs' assessments, and also the program chairs' global view of all papers. The authors of the selected papers were then invited to submit a revised and extended version of their papers having at least 30% innovative material. The second phase started after the receipt of the extended papers. They were reviewed by members of the Program Committee, with emphasis on the scientific value of the augmented innovative material. Moreover, the size of the new material was critically scrutinized (on possessing at least 30% novel material).

The purpose of the book is to disseminate valuable scientific contributions to the field of agent technology and artificial intelligence. The International Conference on Agents and Artificial intelligence (ICAART) is an excellent means to serve this purpose by bringing together researchers, engineers, and practitioners interested in the theory and applications in these areas. Two simultaneous related tracks were held at the conference, covering both agent applications and front-ranked AI research work. The first track focused on agents, multi-agent systems, software platforms, distributed problem solving, and distributed AI in general. The second track focused mainly on artificial intelligence, knowledge representation, planning, learning, scheduling, perception, reactive AI systems, evolutionary computing, and other topics related to intelligent systems and computational intelligence.

The book consists of 11 contributions of high quality. For a list of the papers, we refer readers to the Table of Contents. The proceedings of ICAART 2017 cover the following topics: Knowledge Representation and Reasoning (3), Natural Language Processing (3), Planning and Scheduling (2), Multi-Agent Systems (1), Semantic Web (1), and Intelligent User Interfaces (1).

We would like to thank all the authors for their contributions and also the reviewers, who helped ensure the quality of this publication, for their constructive reports.

February 2017

Jaap van den Herik
Ana Paula Rocha
Joaquim Filipe

Organization

Conference Chair

Joaquim Filipe   Polytechnic Institute of Setúbal/INSTICC, Portugal

Program Co-chairs

Jaap van den Herik   Leiden University, The Netherlands
Ana Paula Rocha   LIACC/FEUP, Portugal

Program Committee

Varol Akman	Bilkent University, Turkey
Isabel Machado Alexandre	Instituto Universitário de Lisboa (ISCTE-IUL) and Instituto de Telecomunicações, Portugal
Vicki Allan	Utah State University, USA
Frédéric Amblard	IRIT, Université Toulouse 1 Capitole, France
Andreas S. Andreou	Cyprus University of Technology, Cyprus
Diana Arellano	Filmakademie Baden-Württemberg, Germany
Tsz-Chiu Au	Ulsan National Institute of Science and Technology, South Korea
Jean-Michel Auberlet	French Institute of Science and Technology for Transport, Development and Networks, France
Kerstin Bach	Norwegian University of Science and Technology, Norway
Federico Barber	Universidad Politécnica de Valencia, Spain
Suzanne Barber	University of Texas, USA
John Barnden	University of Birmingham, UK
Senén Barro	University of Santiago de Compostela, Spain
Roman Barták	Charles University in Prague, Czech Republic
Teresa M. A. Basile	Università degli Studi di Bari, Italy
Punam Bedi	University of Delhi, India
Carole Bernon	University of Toulouse III, France
Daniel Berrar	Tokyo Institute of Technology, Japan
El Hassan Bezzazi	Faculté Droit Lille, France
Elise Bonzon	Lipade, France
Ludovico Boratto	Eurecat, Spain
Rafael Bordini	PUCRS, Brazil
Marco Botta	Università degli Studi di Torino, Italy
Lars Braubach	Universität Hamburg, Germany
Ramón F. Brena	Tecnológico de Monterrey, Mexico
Paolo Bresciani	Fondazione Bruno Kessler, Italy

Muhammad Marwan Muhammad Fuad	Technical University of Denmark, Denmark
Katsuhide Fujita	Tokyo University of Agriculture and Technology, Japan
Naoki Fukuta	Shizuoka University, Japan
Catherine Garbay	CNRS, France
Leonardo Garrido	Tecnológico de Monterrey, Campus Monterrey, Mexico
Alfredo Garro	Università della Calabria, Italy
Benoit Gaudou	University of Toulouse 1 Capitole, France
Jean-Pierre Georgé	University of Toulouse, IRIT, France
Perry Groot	Radboud University Nijmegen, The Netherlands
James Harland	RMIT University, Australia
Hisashi Hayashi	Advanced Institute of Industrial Technology, Japan
Emma Hayes	University of the Pacific, USA
Pedro Rangel Henriques	University of Minho, Portugal
Jaap van den Herik	Leiden University, The Netherlands
Vincent Hilaire	UTBM, France
Hanno Hildmann	Universidad Carlos III de Madrid, Spain
Joerg Hoffmann	Saarland University, Germany
Rolf Hoffmann	Darmstadt University of Technology, Germany
Wladyslaw Homenda	Warsaw University of Technology, Poland
Wei-Chiang Hong	Jiangsu Normal University, China
Mark Hoogendoorn	Vrije Universiteit Amsterdam, The Netherlands
Ales Horak	Masaryk University, Czech Republic
Marc-Philippe Huget	University of Savoie Mont-Blanc, France
Luke Hunsberger	Vassar College, USA
Dieter Hutter	German Research Centre for Artificial Intelligence, Germany
Carlos Iglesias	Universidad Politécnica de Madrid, Spain
Hiroyuki Iida	JAIST, Japan
Jun-ichi Imai	Chiba Institute of Technology, Japan
Thomas Ioerger	Texas A&M University, USA
Luis Iribarne	University of Almería, Spain
Ozgur Kafali	North Carolina State University, USA
Habib M. Kammoun	REGIM-Lab., Tunisia
Geylani Kardas	Ege University International Computer Institute, Turkey
Petros Kefalas	CITY College, International Faculty of the University of Sheffield, Greece
Gabriele Kern-Isberner	TU Dortmund University, Germany
Sung-Dong Kim	Hansung University, South Korea
Matthias Klusch	German Research Center for Artificial Intelligence, Germany
Ah-Lian Kor	Leeds Beckett University, UK
Hristo Koshutanski	Atos Spain, Spain

Andrew Koster	Samsung, Brazil
Igor Kotenko	St. Petersburg Institute for Informatics and Automation of the Russian Academy of Sciences, Russian Federation
Pavel Kral	University of West Bohemia, Czech Republic
Amruth N. Kumar	Ramapo College of New Jersey, USA
Setsuya Kurahashi	University of Tsukuba, Japan
Ramoni Lasisi	Virginia Military Institute, USA
Egons Lavendelis	Riga Technical University, Latvia
Letizia Leonardi	Università di Modena e Reggio Emilia, Italy
Churn-Jung Liau	Academia Sinica, Taiwan
Francesca Alessandra Lisi	Università degli Studi di Bari Aldo Moro, Italy
Juan Liu	Wuhan University, China
Stephane Loiseau	LERIA, University of Angers, France
António Lopes	ISCTE-Instituto Universitário de Lisboa, Portugal
Manuel López-Ibáñez	University of Manchester, UK
Emiliano Lorini	IRIT, CNRS, France
Bernd Ludwig	University of Regensburg, Germany
Daniela Lopéz De Luise	CIIS Lab, Argentina
Letizia Marchegiani	Oxford University, UK
Jerusa Marchi	Universidade Federal de Santa Catarina, Brazil
Philippe Mathieu	University Lille 1, France
Eric Matson	Purdue University, USA
Nicola Di Mauro	Università di Bari, Italy
Francisco José Domínguez Mayo	University of Seville, Spain
Paola Mello	Università di Bologna, Italy
Marjan Mernik	University of Maribor, Slovenia
Elena Messina	National Institute of Standards and Technology, USA
Loizos Michael	Open University of Cyprus, Cyprus
Ambra Molesini	Università di Bologna, Italy
Raul Monroy	Tec de Monterrey in Mexico, Mexico
José Moreira	Universidade de Aveiro, Portugal
Maxime Morge	Université de Lille, France
Bernard Moulin	Université Laval, Canada
Juan Carlos Nieves	Umeå Universitet, Sweden
Jens Nimis	Hochschule Karlsruhe, Technik und Wirtschaft, Germany
Toyoaki Nishida	Kyoto University, Japan
Farid Nouioua	Aix-Marseille University, France
Paulo Novais	Universidade do Minho, Portugal
Luis Nunes	Instituto Universitário de Lisboa and Instituto de Telecomunicações, Portugal
Andreas Oberweis	Karlsruhe Institute of Technology, Germany
Michel Occello	Université Grenoble Alpes, France
Akihiko Ohsuga	University of Electro-Communications, Japan

Haldur Õim	University of Tartu, Estonia
Sancho Oliveira	Instituto Universitário de Lisboa, Portugal
Stanisław Osowski	Warsaw University of Technology, Poland
Nandan Parameswaran	University of New South Wales, Australia
Andrew Parkes	University of Nottingham, UK
Krzysztof Patan	University of Zielona Gora, Poland
Taillandier Patrick	Université de Rouen, UMR IDEES, France
Manuel G. Penedo	University of A Coruña, Spain
Célia da Costa Pereira	Université de Nice Sophia Antipolis, France
Aske Plaat	Tilburg University, The Netherlands
Agostino Poggi	University of Parma, Italy
Enrico Pontelli	New Mexico State University, USA
Filipe Portela	University of Minho, Portugal
Roberto Posenato	Università degli Studi di Verona, Italy
Mariachiara Puviani	Università di Modena e Reggio Emilia, Italy
David Pynadath	University of Southern California, USA
Riccardo Rasconi	National Research Council of Italy, Italy
Marcello Restelli	Politecnico di Milano, Italy
Patrizia Ribino	ICAR, CNR, Italy
Alessandro Ricci	Università di Bologna, Italy
Mark Roberts	Naval Research Laboratory, USA
Ana Paula Rocha	LIACC/FEUP, Portugal
Fátima Rodrigues	Instituto Superior de Engenharia do Porto, Portugal
Daniel Rodriguez	University of Alcalá, Spain
Irene Barba Rodriguez	Universidad de Sevilla, Spain
Andrea Roli	Università di Bologna, Italy
Juha Röning	University of Oulu, Finland
Javier Carbó Rubiera	Universidad Carlos III de Madrid, Spain
Alvaro Rubio-Largo	University of Extremadura, Spain
Ruben Ruiz	Universidad Politécnica de Valencia, Spain
Luca Sabatucci	ICAR-CNR, Italy
Fariba Sadri	Imperial College London, UK
Francesco Santini	Università di Perugia, Italy
Jorge Gomez Sanz	Universidad Complutense de Madrid, Spain
Fabio Sartori	Università degli Studi di Milano Bicocca, Italy
Jurek Sasiadek	Carleton University, Canada
Francesco Scarcello	Università della Calabria, Italy
Stefan Schiffer	RWTH Aachen University, Germany
Christoph Schommer	University of Luxembourg, Luxembourg
Stefan Schulz	Medical University of Graz, Austria
Michael Schumacher	University of Applied Sciences and Arts Western Switzerland, Switzerland
Martijn Schut	University of Amsterdam, The Netherlands
Frank Schweitzer	ETH Zurich, Switzerland
Valeria Seidita	University of Palermo, Italy
Emilio Serrano	Universidad Politécnica de Madrid, Spain

Feiyu Xu	Deutsches Forschungszentrum für Künstliche Intelligenz, Germany
Neil Yorke-Smith	TU Delft, The Netherlands
Jing Zhao	ECNU, China

Additional Reviewers

Syed Saqib Bukhari	German Research Center for Artificial Intelligence, Germany
Alevtina Dubovitskaya	HES-SO Valais; EPFL, Switzerland
Francesco Lupia	University of Calabria, Italy
Pablo Pilotti	Cifasis, Argentina
Christos Rodosthenous	Open University of Cyprus, Cyprus
Yuichi Sei	University of Electro-Communications, Japan
Razieh Nokhbeh Zaeem	UT Austin, USA
Matteo Zavatteri	University of Verona, Italy

Invited Speakers

Vanessa Evers	University of Twente, The Netherlands
João Leite	Universidade Nova de Lisboa, Portugal
Nuno Lau	Universidade de Aveiro, Portugal
Francesca Rossi	IBM, USA and University of Padua, Italy

Contents

A Description Logic Based Knowledge Representation Model
for Concept Understanding. 1
   Farshad Badie

Reasoning for Autonomous Agents in Dynamic Domains:
Towards Automatic Satisfaction of the Module Property 22
   Stephan Opfer, Stefan Jakob, and Kurt Geihs

Chest X-Ray Image Analysis: A Vision of Logic Programming 48
   João Neves, Ricardo Faria, Victor Alves, Filipa Ferraz,
   Henrique Vicente, and José Neves

Text Classification and Transfer Learning Based on Character-Level
Deep Convolutional Neural Networks 62
   Minato Sato, Ryohei Orihara, Yuichi Sei, Yasuyuki Tahara,
   and Akihiko Ohsuga

A Hierarchical Playscript Representation of Distributed Words
for Effective Semantic Clustering and Search 82
   Avi Bleiweiss

Data Harvesting and Event Detection from Czech Twitter 102
   Václav Rajtmajer and Pavel Král

Variants of Independence Detection in SAT-Based Optimal
Multi-agent Path Finding 116
   Pavel Surynek, Jiří Švancara, Ariel Felner, and Eli Boyarski

ε-Strong Privacy Preserving Multi-agent Planning 137
   Antonín Komenda, Jan Tožička, and Michal Štolba

A Quantitative Assessment of the JADEL Programming Language 157
   Federico Bergenti, Eleonora Iotti, Stefania Monica, and Agostino Poggi

Analogical Reasoning in Clinical Practice with Description Logic $\mathcal{ELH}$. 179
   Teeradaj Racharak and Satoshi Tojo

Advanced User Interfaces for Semantic Annotation of Complex
Relations in Text. 205
   Jaroslav Dytrych and Pavel Smrz

Author Index 223

A Description Logic Based Knowledge Representation Model for Concept Understanding

Farshad Badie[(⊠)]

Center for Computer-mediated Epistemology, Aalborg University,
Rendsburggade 14, 9000 Aalborg, Denmark
badie@id.aau.dk

Abstract. This research employs Description Logics in order to focus on logical description and analysis of the phenomenon of 'concept understanding'. The article will deal with a formal-semantic model for figuring out the underlying logical assumptions of 'concept understanding' in knowledge representation systems. In other words, it attempts to describe a theoretical model for concept understanding and to reflect the phenomenon of 'concept understanding' in terminological knowledge representation systems. Finally, it will design an ontology that schemes the structure of concept understanding based on the proposed semantic model.

Keywords: Concept understanding · Conceptualisation
Terminological knowledge · Interpretation · Formal semantics
Description logics · Ontology

1 Introduction and Motivation

My point of departure is the special focus on the fact that there has always been a general problem concerning the notion of 'concept', in linguistics, psychology, philosophy, and computer science. In Kant's opinion, a concept is the "unity of the act of bringing various representations under one common representation". In addition, Kant believed that "no concept is related to an object immediately, but only to some other representation of it", see [16]. However, it has never been transparent (i) if concepts are some mental representations as well as mental images of various phenomena, or (ii) whether concepts always have to be bound up (and thus, be labelled) with some linguistic expressions.

Concepts are the main building blocks of this research. Note that this article, as an extended version of [4], aims at providing a logical analysis of a specific use of the phenomenon of 'concept' in terminological knowledge representation systems. In order to see and conceptualise concepts from the perspective of logics, we need to interpret them some logical-assessable phenomena. For example, we can interpret a concept a set (class) like the set of *Tree*s. Consequently, a set, that is an understandable and assessable mathematical phenomenon, can be applied to different contexts. Accordingly, a concept might be considered a conceptual entity and, in fact, could be

© Springer International Publishing AG, part of Springer Nature 2018
J. van den Herik et al. (Eds.): ICAART 2017, LNAI 10839, pp. 1–21, 2018.
https://doi.org/10.1007/978-3-319-93581-2_1

correlated with a distinct 'entity' or with its essential attributes, characteristics, and properties. Note that a conceptual entity's properties express its relationships with itself and with other conceptual entities (e.g., reflexive, irreflexive, symmetrical, anti-symmetrical, transitive, anti-transitive relationships).

Through the lens of First-Order Predicate Logic, an entity is assessable a [unary] predicate. Accordingly, such an equivalence between a [conceptual] entity and a unary predicate can support terminological descriptions as well as logical representations of that conceptual entity. The main task of predicates is 'assigning'. In fact, predicates make mental mappings from the attributes and properties of conceptual entities into subjects, see 'predicate' in [7]. Furthermore, predicates can express the conditions that the conceptual entities referred to may satisfy. So, the most central logical premise is that predicates describe conceptual entities in order to determine the applications of logical descriptions and, accordingly, to play fundamental roles in reasoning processes and in giving satisfying conditions for definitions of truth. Consequently, predicates, as the outcomes of predications, express meanings and produce formal semantics. Subsequently, a formal semantics could focus on multiple conditions through definitions of truth (and falsity). Note that any formal semantics deals with the interrelationships between the signifiers of a description and what the signifiers do [or have been designed to do], see [5, 12, 15, 20]. This could be interpreted the most significant essence of any formal semantics.

In this research, concepts (conceptual entities) and their interrelationships will be employed to establish the basic terminology adopted in the modelled domain regarding the hierarchical structures. Relying on such a hierarchical structure, this research focuses on logical description as well as logical analysis of the phenomenon of 'concept understanding'. The desired logical descriptions will have a special focus on my methodological assumption that expresses that "one can find out that an individual thing/phenomenon is an instance of a concept (conceptual entity) and, thus, his[1] individual grasp of that concept (in the form of his conceptions) provide supportive foundations for producing his own conceptualisations. Accordingly, he restricts his produced conceptualisation to his concept understanding". This article will focus on describing and characterising humans' concept understandings and will deal with a formal-semantic model for uncovering the underlying logical assumptions of 'concept understanding' in knowledge representation systems. In other words, it attempts to describe a theoretical model for concept understanding and to reflect it in termino-logical knowledge representation systems. In this research, the phenomenon of 'con-cept understanding' will be seen from multiple perspectives. Subsequently, the expressiveness (as well as complexity) of the desired semantic model's descriptions will be improved (and increased).

In this research, the formal semantic analysis of concept understanding is based on Description Logics (DLs). DLs can support me in proposing a comprehensible logical description for clarifying the phenomenon of 'concept understanding'. DLs are, as the most well-known knowledge representation formalisms, used for representing predi-cates and for formal reasoning over them. They mainly focus on terminological

---

[1] For brevity, I use 'he' and 'his' whenever 'he or she' and 'his or her' are meant.

knowledge. It is of a terminological system's particular importance in providing a logical formalism for knowledge representation systems and, also, for ontology representations. In information and computer sciences, ontologies are formal and explicit specification of a shared conceptualisation, see [9, 23]. This research, thus, will focus on building up as well as formalising an ontology for 'concept understanding'. The desired ontology provides a structural representation of concept understanding based on the analysed semantic model.

2 The Phenomenon of 'Concept Understanding'

The term 'understanding' is very complicated and sensitive in psychology, neuroscience, cognitive science, philosophy, and epistemology. It shall be emphasised that there has not been any adequate model for understanding. More specifically, there has not been any complete, deterministic, and unexceptionable model for describing the phenomenon of 'understanding'. Anyhow, there have been some proper models for:

1. understanding of understanding (e.g., [11]),
2. understanding representation (e.g., [19, 26]), and
3. specification of the components of understanding (i.e., from the cognition's as well as desires' and emotions' perspectives), e.g., [8, 10, 17, 18, 24, 25, 27].

The first category of models can describe how the phenomenon of 'understanding' could be realised and figured out in different contexts. The second category focuses on epitomising, designing, visualising, and illustrating the phenomenon of 'understanding' in different contexts. And finally, the third category focuses on recognising and specifying the most significant ingredients and constructors of the phenomenon of 'understanding' (mostly from the cognitive perspectives).

Taking into consideration the phenomenon of 'understanding', the following assumptions can describe my main conceptions of the phenomenon of 'concept understanding':

1. I shall assume that if one is going to produce his understanding based on a concept (conceptual entity), then, he will be, either directly or indirectly, become concerned with the ontology as well as the existence of that concept. For instance, producing understanding based on the entity 'Tree' makes one concerned with the existence and nature of the concept 'Tree'. Therefore, it must be accepted that the phenomenon of 'concept understanding' relates a human being with the existence of a conceptual entity.
2. There is a strong interconnection and dependency between the phenomenon of 'concept understanding' and the phenomenon of 'explanation'. In addition, it shall be taken into account that the conceptual relationships between the explanans (that which does the explaining) and the explanandum (that which is to be explained) support the phenomenon of 'concept understanding'. Consequently, the phenomenon of 'explanation', as the outcome of the logical relationships between the explanans and the explanandum, attempts to shed light on the concept understanding's targets. For example, explaining Tree and understanding Tree are strongly interrelated to each other.

3. In order to describe the phenomenon of 'concept understanding', we can see it from the structuralist point of view. Such an overview can support us in explaining different facts and procedures about 'concept understanding'. It is undeniable that understanding *Tree* and understanding *Bough* are tied to each other. Obviously, observing 'concept understanding' from the structuralist point of view can connect us to the dependencies between the phenomenon of 'concept understanding' and the phenomenon of 'explanation'. Accordingly, we may become concerned with two issues.

   a. With the taxonomies and hierarchies of concept understanding. This means that seeing concept understanding from the structuralist point of view can link us to the existence of the phenomenon of 'concept understanding' and, subsequently, to concept understanding's different levels of complexity. For example, understanding *Tree* can, inductively, connect us with understanding 'young *Tree*' and understanding 'flowers of a young *Tree*'.

   b. With multiple aspects/criteria of concept understanding (Specially from the perspective of Cognition) that can relate various components of concept understanding to each other. Note that we may also apply different tools (e.g., graphical tools, statistical tools, concept maps) in order to, inductively, relate different components of a concept understanding to each other.

   In Sect. 6, you will see that my semantic model focuses on the junctions between 'understanding of concept understanding' and 'concept understanding representation'.

3   Description Logics

My main reference to Description Logics is [1]. Description Logics (DLs) represent knowledge in terms of (i) *individuals* (objects, things), (ii) *concepts* (classes of individuals), and (iii) *roles* (relationships between individuals). Individuals correspond to constant symbols, concepts to unary predicates, and roles to binary (or any other n-ary) predicates in Predicate Logic. A predicate in Predicate Logic can have a [possibly specified] equivalent concept in DLs. There are two kinds of atomic symbols, which are called *atomic concepts* and *atomic roles*. These symbols are the elementary descriptions from which we can inductively (by employing concept constructors and role constructors) construct the specified descriptions. Considering N_C, N_R, and N_O the sets of atomic concepts, atomic roles, and individuals, respectively the ordered triple (N_C, N_R, N_O) denotes a signature. The set of main logical symbols in ALC (Attributive Concept Language with Complements) that is the Prototypical DL (see [21]) is:

$$\{\text{Conjunction}(\sqcap : \text{And}), \text{Disjunction}(\sqcup : \text{Or}), \text{Negation}(\neg : \text{Not}),$$
$$\text{Existential Restriction}(\exists : \text{There exists}\ldots), \text{Universal Quantification}(\forall : \text{For all}\ldots)\}.$$

In addition, ALC contains:

$$\{\text{Atomic Concepts}(A),$$
$$\text{Top Concept}(\top : \text{Tautology}), \text{Bottom Concept}(\bot : \text{Contradiction})\}$$

In order to define a formal semantics, we need to apply terminological interpretations over our signatures. More particularly, any [terminological] interpretation consists of:

(a) a non-empty set Δ that is the interpretation domain and consists of any variable that occurs in any possible concept description, and
(b) an interpretation function $(.^I)$. I prefer to name it 'interpreter'.

The interpreter assigns m^I to every individual m. Note that m^I is in Δ^I (i.e., $m^I \in \Delta^I$). Also, it assigns to every atomic concept A, a set $A^I \subseteq \Delta^I$. In addition, it assigns to every atomic role P (between two individuals), a binary relation like P^I, such that: $P^I \subseteq \Delta^I \times \Delta^I$. This relationship is inductively extendable for any n-ary relationship over the interpretation domain. Table 1 reports the syntax and the semantics of ALC.

Table 1. The prototypical description logic

Syntax	Semantics
A	$A^I \subseteq \Delta^I$
P	$P^I \subseteq \Delta^I \times \Delta^I$
$\top$	Δ^I
$\bot$	$\emptyset$
$C \sqcap D$	$(C \sqcap D)^I = C^I \cap D^I$
$C \sqcup D$	$(C \sqcup D)^I = C^I \cup D^I$
$\neg C$	$(\neg C)^I = \Delta^I \setminus C^I$
$\exists R.\ C$	$\{a \mid \exists b.(a,b) \in R^I \wedge b \in C^I\}$
$\forall R.\ C$	$\{a \mid \forall b.(a,b) \in R^I \supset b \in C^I\}$

A knowledge base in DLs usually consists of a number of terminological axioms and world descriptions (so-called 'assertions'), see Table 2.

Table 2. Axioms and world descriptions in DLs

Name	Syntax	Semantics
Concept inclusion axiom	$C \sqsubseteq D$	$C^I \subseteq D^I$
Role inclusion axiom	$R \sqsubseteq S$	$R^I \subseteq S^I$
Concept equality axiom	$C \equiv D$	$C^I = D^I$
Role equality axiom	$R \equiv S$	$R^I = S^I$
Concept assertion	$C(a)$	$a^I \in C^I$
Role assertion	$R(a, b)$	$(a^I, b^I) \in R^I$

In DLs, in case a given terminological interpretation could assign the value True to a concept description, that interpretation is called a 'model' of that description. Consequently, a terminological interpretation (like I) can be a model of a terminological and, respectively, of an assertional description if and only if it can satisfy them semantically, see Tables 2 and 3. In these Tables P is an atomic role, R and S are role descriptions, A is an atomic concept, and C and D are concept descriptions.

Table 3. Inductive concept descriptions.

Over concept	Over role
$A^I \subseteq \Delta^I$	$P^I \subseteq \Delta^I \times \Delta^I$
$\perp^I = \varnothing$	$\perp^I = \varnothing$
$(\neg C)^I = \Delta^I \setminus C^I$	$(\neg R)^I = (\Delta^I \times \Delta^I) \setminus R^I$
$(C \sqcap D)^I = C^I \cap D^I$	$(R \sqcap S)^I = R^I \cap S^I$

4 Logical Characterisation of Terminological Knowledge

Description Logics (DLs) are a family of semi-formal descriptive languages. Any DL (like ALC) represents concepts and their interrelationships in order to represent terminological knowledge. Subsequently, it provides a logical backbone for concept-based reasoning processes. In this section, I do focus on logical analysis of the terminological background of concept-based reasoning.

Suppose that the function $M_K(C)$ denotes that machine (M) has—on a basis supported by its terminological knowledge (K)—focused on the concept C. Let T and W stand for a terminology and a world description, respectively. Thus, the terminological knowledge is equal to (T, W). More specifically, considering E_C as a set of examples of the concept C,

$$E_C = \{E_C^+, E_C^-\}$$

where, E_C^+ and E_C^- stand for positive and negative examples of C, respectively. Consequently:

- If a concept assertion (like $D(a)$) is satisfied by K (and, in fact, by (T, W)), then:

$$\forall a \in E_C^+(W), K \models D(a).$$

- If a concept assertion (like $D(b)$) is not satisfied by K, then:

$$\forall b \in E_C^-(W), K \nvDash D(b).$$

The logical symbol '$\models$' in the term '$K \models D(a)$' denotes that knowledge K has been supportive (and satisfactory) for satisfying the concept description $D(a)$. In fact, K (as a collection of terminology T and world description W) can satisfy the expressed concept description. Also, the symbol '$\nvDash$' in the term '$K \nvDash D(b)$' describes that K has not been supportive (and satisfactory) for satisfying the concept description $D(b)$.

These conclusions are also valid in the case of roles. More specifically, $M_K(R)$ denotes that machine M has—on a basis supported by its terminological knowledge (K)—focused on the n-ary role (R). Consequently:

$$E_R = \{E_R^+, E_R^-\}.$$

Therefore:

- If the role assertion $R(a_1, a_2, ..., a_n)$ is satisfied by K (and, in fact, by (T, W)), then:

$$\forall a_1, a_2, \ldots, a_n \in E_R^+(W), K \vDash R(a_1, a_2, \ldots, a_n).$$

- If the role assertion $R(b_1, b_2, ..., b_n)$ is not satisfied by K, then:

$$\forall b_1, b_2, \ldots, b_n \in E_R^-(W), K \nvDash R(b_1, b_2, \ldots, b_n).$$

5 Logical Clarification of Concept Understanding

This section offers two examples in order to deal with a logical clarification of the phenomenon of 'concept understanding'.

5.1 Example I

Mary thinks that the term 'there is a young student' and the term 'there is a non-old student' are equivalent to each other. Mary's verification between these two propositions is expressible in DLs by:

$$\exists hasStudent.Young \equiv \exists hasStudent.\neg Old.$$

We can figure out that Mary has, mentally, assumed the axiom stating that Young and Old are two disjoint (= distinct) concepts. In fact, the logical description 'Young $\sqcap$ Old $\sqsubseteq \bot$' has formed a presupposition (in the form of a terminological axiom) in Mary's mind. It's obvious that Mary's interpretation has played crucial roles here. More specifically, her terminological interpretation has been carried out based on the following fundamental logical descriptions:

i. Young $\sqcap$ Old $\sqsubseteq \bot$. This fundamental description expresses that Young and Old are two disjoint concepts in Mary's mind.
ii. Person $\sqsubseteq$ Young $\sqcup$ Old. This fundamental description means that every person is either young or old in Mary's mind. Equivalently, any person could be described (and predicated) either by the predicate Young or by the predicate Old.

Mary has interpreted and, respectively, has understood that the proposition 'there is a young student' and the proposition 'there is a non-old student' have the same meanings. More specifically, Mary's terminological interpretations (over 'Young $\sqcap$ Old $\sqsubseteq \bot$' and 'Person $\sqsubseteq$ Young $\sqcup$ Old') have produced her understanding of the equivalence ($\equiv$) between the concept descriptions '$\exists hasStudent.Young$' and '$\exists hasStudent.\neg Old$'. We can see that Mary's interpretation has been restricted (limited) to her understanding of the disjointness of the concept descriptions '$\exists hasStudent.Young$' and

'∃hasStudent.¬Old'. Note that two concepts (concept descriptions) like C and D are logically and semantically equivalent when, 'for all' possible terminological interpretations like I, we have: $C^I = D^I$.

If one other person, say John, does not assume the axioms stating that 'Young and Old are two disjoint concepts' and 'every person is either young or old', then there will not be an equivalence relation between ∃hasStudent.Young and ∃hasStudent.¬Old. Let me conclude that Mary's and John's concept understandings are dissimilar, because they have had different terminological interpretations in their minds. Such a difference is caused by their different conceptions of the world. For example, John may—regarding his terminological interpretation—believe that the proposition 'there is a middle-aged student' comes next to 'there is a young student' and 'there is a non-old student'. In fact, John keeps in mind the axiom 'Person ⊑ Young ⊔ MiddleAged ⊔ Old'. It means that every person is young or middle-aged or old. Consequently, John by taking this axiom (based on his own conception) into consideration doesn't understand '∃hasStudent.Young' and '∃hasStudent.¬Old' as equivalent concept descriptions.

5.2 Example II

Mary believes that the propositions 'Anna has a child who is a philosopher' and 'Anna has a child who is a painter' could be jointly expressed by the proposition 'Anna has a child who is a philosopher and painter'. Translated into DLs we have her description as:

$$(\exists\, hasChild.Philosopher \sqcap \exists\, hasChild.Painter) \equiv$$
$$\exists hasChild.(Philosopher \sqcap Painter).$$

Suppose that Anna has two children and one is a philosopher and the other one is a painter. Then, ∃hasChild.(Philosopher ⊓ Painter) is not equivalent to ∃hasChild.Philosopher ⊓ ∃hasChild.Painter, because the one who is a philosopher, is not a painter, and vice-versa.

Actually, Mary has not proposed a correct description. Her non-correct description is caused by her inappropriate terminological interpretation. Accordingly, her concept understanding has followed her inappropriate interpretation. In fact, she has incorrectly understood that the proposition 'Anna has a child who is a philosopher and painter' is equivalent to the collection of the propositions 'Anna has a child who is a philosopher' and 'Anna has a child who is a painter'. Reconsidering the proposed formalism, the concept descriptions:

1. '∃hasChild.Philosopher ⊓ ∃hasChild.Painter', and
2. '∃hasChild.(Philosopher ⊓ Painter)'

are not, semantically, the same. In fact, there should not be an equivalence symbol between them. Thus, Mary's interpretation has not been satisfactory. Subsequently, her concept understanding is not satisfactory and appropriate.

6 A Semantic Model for Concept Understanding

Relying on Sect. 4, this section focuses on logical analysis of concept understanding and its terminological representation. More specifically, this section by taking into account my logical conceptions of 'concept understanding' (offered in Sect. 2) analyses a formal semantics and, subsequently, focuses on the junctions between 'understanding of concept understanding' and 'concept understanding representation' in terminological systems.

6.1 Concept Understanding as a Relation (and Function)

I shall claim that 'concept understanding' expresses a relationship. This relationship relates 'the characteristics and attributes of a concept' to 'a description'. More specifically, *understanding* is a function (mapping) from a concept (conceptual entity) as well as its attributes into a statement. In fact, one could, based on his personal concept understanding, propose his personal concept descriptions. Therefore:

$$Concept\ Understanding : Concept \rightarrow Concept\ Description.$$

Let me be more specific:

A. A human being—by concept understanding—attempts to map the significant characteristics of concepts into some concept descriptions. For example, 'breathing', as a biological and psychological process, is a characteristic and trait of all animals. Then, breathing (that is a role) is the characteristic of the concept Animal. Therefore, (i) knowing the fact that the individual 'horse' is an instance of the concept 'Animal' (Formally: Animal(horse)), and (ii) drawing the [concept subsumption] inference 'Horse ⊑ Animal', collectively lead us to knowing and, subsequently, to understanding that 'horses breathe' (or equivalently: 'horses do breathing'). Note that the role 'breathing' could be manifested in the concept 'Breath'. Therefore, (i) and (ii) collectively lead us to expressing the concept description 'Animal(horse) ⊓ ∃hasTrait.Breath' for the individual 'horse' (as an instance of the concept 'Animal') and, respectively, for the concept 'Horse' (as a sub-concept of 'Animal').

B. A human being—by concept understanding—attempts to map the concepts' reflexive properties (concepts' interrelationships with themselves) into some concept descriptions. For example, one who knows that 'male horses breathe', by taking into consideration the terminological and assertional axioms:

{
Animal(horse),
Horse ⊑ Animal,
MaleHorse ⊑ Horse,
FemaleHorse ⊑ Horse
},

can figure out and, accordingly, can understand that 'female horses breathe' as well.

C. A human being—by concept understanding—attempts to map the concepts' properties (and their relationships with other concepts) into some concept descriptions. For example, one who knows that 'horses breathe' (and as described: Animal(horse) ⊓ ∃hasTrait.Breath), could, respectively figure out and understand that the individual 'rabbit' (that is an Animal) breathes as well. So, he could express that 'rabbits breathe' and, in fact, Animal(rabbit) ⊓ ∃hasTrait.Breath.

Conclusion. Relying on Predicate Logic (and on DLs), the phenomenon of 'concept understanding' could be interpreted a 'binary predicate' (and a 'role' of human beings on expressing concept descriptions). This role will be represented by 'understanding' in my formalism.

6.2   Concept Understanding as a Conceptualisation

Concept understanding could be interpreted the limit/type of conceptualisation. Accordingly, humans need to conceptualise concepts in order to understand them. In [2, 3], I have interpreted a 'concept understanding' a local manifestation of a global 'conceptualisation'. Then, I acknowledge one's 'concept understanding' as a limited type of his own conceptualising. Note that 'conceptualising' could be recognised his role. This conclusion, relying on DLs, could be represented by the role inclusion (or role sub-sumption):

$$\text{understanding} \sqsubseteq \text{conceptualising}.$$

In fact, considering C a concept,

$$understanding \; C \subseteq conceptualising \; C.$$

On the other hand, 'it is not the case that all conceptualisations are concept understandings'. In fact, all the conceptualised concepts are not necessarily understood.

6.3   Concept Understanding as an Interpretation-Based Model

Generally, an interpretation is the act of elucidation, explication, and explanation, see [22]. According to [14] and through the lens of philosophy, "...in existential and hermeneutic philosophy, interpretation becomes the most essential moment of human life: The human being is characterized by having an 'understanding' of itself, the world, and others. This understanding, to be sure, does not consist—as in classical ontology or epistemology—in universal features of universe or mind, but in subjective–relative and historically situated interpretations of the social. ...". Regarding [7] and through the lens of logic, an 'interpretation' of a logical system assigns meanings as well as semantic values to the formulae and their elements. At this point I shall emphasise that formal languages may see the phenomenon of 'interpretation' terminologically. In fact, a logical-terminological system can restrict the phenomenon of 'interpretation' to the phenomenon of 'terminological interpretation' in order to assess and apply it in logical as well as terminological contexts. More specifically, from the perspective of a logical-terminological system, one who has engaged his interpretations to explicate [and justify] what [and why] he means by classifying a thing/phenomenon as an instance of a concept, needs to interpret the non-logical signifiers of various concept descriptions within his linguistic expressions.

Considering any set of non-logical symbols (that have no logical consequences) in a terminology, a terminological interpretation of humans' languages could be described to be constructed based on the tuple:

$$(Interpretation\ Domain, Interpretation\ Function).$$

The first component (the interpretation domain) expresses the universe of the interpretation. Note that in linguistic and philosophical approaches it might be called 'universe of discourse'.

As mentioned above, an interpretation domain (like D) must be non-empty (i.e.., $D \neq \varnothing$). This non-empty set supports the range of any variable that occurs in any of the concept descriptions within logical descriptions of linguistic expressions. It is a fact that the collection of the rules and the processes that manage different terms and logical descriptions in linguistic expressions, cannot have any meaning until the non-logical signifiers and constructors are given terminological interpretations. The interpretations prepare human beings for producing their personal meaningful as well as under-standable concept descriptions. Hence, I believe that all 'concept understandings' are 'concept interpretations'. According to [4], this conclusion could be represented by the role inclusion:

$$understanding \sqsubseteq interpreting.$$

In fact, considering C a concept,

$$understanding\ C \subseteq interpreting\ C.$$

But, on the other hand, all interpretations (over concepts) do not imply under-standings (of concepts). Equivalently, 'it is not the case that all concept interpretations are concept understandings'. In other words, all the interpreted concepts may not be understood. Accordingly, considering any interpretation a function, 'concept under-standing' is recognised an 'interpretation function'.

From this point I apply the function UND (as a limit of the interpretation function I) in my formalism. Then, considering C a concept,

$$C^{UND} = Understanding\ C.$$

Consequently, considering UND a kind of interpretation, there exists a tuple like $(D_U, C_{understood})$, where:

  i. D_U represents the understanding domain (that consists of the variables that occur in any of the concept descriptions that are going to be understood), and

  ii. $C_{understood}$ is the understood concept.

$C_{understood}$ is achievable based on the understanding function $-^{UND}$. Relying on the function $-^{UND}$,

$$C^{UND} \subseteq C^I \subseteq \Delta^I$$
$$\&$$
$$D_U^{UND} \subseteq \Delta^I.$$

D_U^{UND} means 'understanding all concepts belonging to the understanding domain'. Note that $-^{UND}$ (that is a function) can provide a model for terminological and, respectively, for assertional axioms. Therefore, the desired model:

- is a restricted form of a terminological-interpretation-based model, and
- can satisfy the semantics of the terminological and assertional axioms ('$UND \models$ Axiom' expresses that UND satisfies the axiom).

See Table 4. Consequently:

Table 4. *Concept Understanding*: terminologies and world descriptions.

Name	Description and semantics
Understanding a concept inclusion	$[UND \models (C \sqsubseteq D)] \Rightarrow [C^{UND} \subseteq D^{UND}]$
Understanding a role inclusion	$[UND \models (R \sqsubseteq S)] \Rightarrow [R^{UND} \subseteq S^{UND}]$
Understanding a concept equality	$[UND \models (C \equiv D)] \Rightarrow [C^{UND} = D^{UND}]$
Understanding a role equality	$[UND \models (R \equiv S)] \Rightarrow [R^{UND} = S^{UND}]$
Understanding a concept assertion	$[UND \models C(a)] \Rightarrow [a^{UND} \in C^{UND}]$
Understanding a role assertion	$[UND \models R(a_1, a_2, ..., a_n)] \Rightarrow$ $[(a_1^{UND}, a_2^{UND}, ..., a_n^{UND}) \in R^{UND}]$

$$C^{UND} \subseteq C^I \subseteq \Delta^I$$
$$\&$$
$$-^{UND} : C \rightarrow C^{UND}$$
$$\text{Where}: C^{UND} \subseteq D_U^{UND} \subseteq \Delta^I.$$

I shall emphasise that we are not able to conclude that $C^I \subseteq D_U^{UND}$. On the other hand, we certainly know that $C^{UND} \subseteq \Delta^I$ (because $C^{UND} \subseteq C^I$ and $C^I \subseteq \Delta^I$). According to the analysed characteristics, the UND understanding model in my terminology is constructed over the tuple:

$$(Understanding\ Domain, Understanding\ Function).$$

And, formally:

$$UND = (D_U^{UND}, -^{UND}).$$

Table 5 represents understanding inductive concept descriptions as the products of the proposed understanding model. This Table is logically supported by Table 4, see [4].

Table 5. Understanding inductive concept descriptions.

Model satisfies the vocabulary	Semantics
$UND \models \top$	$\top^{UND} = \top$
$UND \models \bot$	$\bot^{UND} = \varnothing$
$UND \models \neg R$	$(\neg R)^{UND} = \top \backslash R^{UND}$
$UND \models \neg C$	$(\neg C)^{UND} = D_U^{UND} \backslash C^{UND}$
$UND \models (R \sqcap S)$	$(R \sqcap S)^{UND} = R^{UND} \cap S^{UND}$
$UND \models (C \sqcap D)$	$(C \sqcap D)^{UND} = C^{UND} \cap D^{UND}$

6.4   Concept Understanding as a Consequence of Functional Roles

How could we employ DLs in order to describe a [concept] understanding function? In my opinion, an understanding function must be interpreted a functional role of human beings in order to be, logically, described. The functional roles (features) are the roles that are structurally as well as inherently functions and, hence, they can express *functional* actions, movements, procedures, and manners of human beings.

Let N_F be a set of functional roles and N_R be the set of roles (role descriptions). Obviously: $N_F \subseteq N_R$. Informally, functional roles are some kinds of roles.

Lemma. The *UND* understanding model is, semantically, structured based on:

a. the understanding domain (or D_U),
b. the understanding function (or $\cdot^{UND}$), and
c. the set D_U^{UND} (or equivalently, the effect of the understanding function $\cdot^{UND}$ on the Top concept) that represents understanding all atomic concepts (everything) in the understanding domain.

Analysis. The *UND* model associates with each atomic concept a subset of D_U^{UND} and with each ordinary atomic role a binary relation over $D_U^{UND} \times D_U^{UND}$. Assessed by Mathematics, any functional role can be seen as a partial function. More specifically, considering F as a chain of functional roles or, equivalently,

$$F = f_1 \circ \cdots \circ f_n,$$

the composition of n partial concept understanding functions can be represented by:

$$f_1^{UND} \circ \cdots \circ f_n^{UND}.$$

In fact, by employing *UND*, any f_i^{UND}—semantically—supports the overall functional role F^{UND}. Note that for all i in $(1,n)$, f_{i+1} produces the input of f_i. Therefore, understanding f_{i+1} (the output of f_{i+1}) provides the input of understanding f_i. In particular, any concept description could be understood over the subsets of D_U^{UND}. This characteristic is very useful in making a strong linkage between the terms 'concept understanding' and 'chain of functional roles'. It supports my semantic model in scheming and describing '*the concept understanding as the product of a chain of functional roles, where the functional roles are the partial understanding functions'.* You will see how it works.

6.5   Humans' Functional Roles Through SOLO's Levels

According to [6], the Structure of Observed Learning Outcomes (SOLO) taxonomy is a proper model that can provide an organised framework for representing different levels of humans' understandings. This model is concerned with various complexities of understanding on its different layers. According to SOLO taxonomy and taking into consideration humans' multiple layers of knowledge (based on concepts), we have:

- Pre-structured knowledge. Here, humans' knowledge of a concept is pre-structured. The pre-structured knowledge is the product of one's pre-conceptions of a concept.
- Uni-structured knowledge. Humans have a limited knowledge about a concept. Having a uni-structured knowledge is the outcome of knowing one or few isolated fact(s) about a concept.
- Multi-structured knowledge. Humans are getting to know a few facts relevant to a concept, but they are still unable to link and relate them together.
- Related Knowledge. Humans have started to move towards deeper levels of understanding of a concept. Here, they are able to explain their several conceptions of a concept. Also, they can link different facts (regarding their conceptions of a concept) to each other.
- Extended Abstracts. This is the most complicated level. Humans are not only able to link lots of related conceptions (of a concept) to each other, but they can also link them to other specified and complicated conceptions. Now, they are able to link multiple facts and explanations in order to produce more complicated extensions relevant to a concept.

Obviously, the extended abstracts are the products of deeper comprehensions of related structures. Related structures are the products of deeper comprehensions of multi-structures. The multi-structures are the products of deeper comprehensions of uni-structures, and the uni-structures are the products of deeper comprehensions of pre-structures.

Let me select a process as a sample of humans' functional roles from any of the SOLO's levels and formalise it. According to SOLO, (a) the phenomenon of *'creation'* (based on a concept) is an instance of the 'extended abstracts', (b) the phenomenon of *'justification'* (based on a concept) is an instance of the 'related structures', (c) the phenomenon of *'description'* (based on a concept) is an instance of the 'multi-structures', and (d) the phenomenon of *'identification'* (based on a concept) is an instance of the 'uni-structures'. Therefore, the phenomena of *'Creation'*, *'Justification'*, *'Description'*, and *'Identification'* are four processes. These processes can be seen and interpreted functions in my semantic model. More specifically, any of these functions can support a functional role and, subsequently, can support a 'partial concept understanding function'. Actually,

i. *Creation* has interrelatedness with creatingOf that is a functional role and extends the humans' mental abstracts.

ii. *Justification* has interrelatedness with the functional role justifyingOf. This functional role relates the lower structures.

iii. *Description* has correlation with the functional role describingOf. This role produces the multi-structures.

iv. *Identification* has correlation with the functional role identifyingOf that generates the uni-structures.

It shall be emphasised that identifyingOf, describingOf, justifyingOf, and creatingOf are only four examples of functional roles within SOLO's categories and, in fact, the SOLO's levels are not limited to these functions. For example, followingOf and namingOf are two other instances of the uni-structures, combiningOf and enumeratingOf are two other instances of the multi-structures, analysingOf and arguingOf are two other instances of the related structures, and formulatingOf and theorisingOf are two other instances of the extended abstracts.

As mentioned, the functional roles creatingOf, justifyingOf, describingOf, and identifyingOf represent the equivalent roles of the *creation, justification, description,* and *identification* functions, respectively. Furthermore, these functions are the partial functions of the [concept] *understanding* function. Obviously, the concept understanding function (that is a process) could also be considered to be equivalent to a functional role like understandingOf. Employing the 'role inclusion' axiom we have:

(1) creatingOf $\sqsubseteq$ understandingOf,
(2) justifyingOf $\sqsubseteq$ understandingOf,
(3) describingOf $\sqsubseteq$ understandingOf, and
(4) identifyingOf $\sqsubseteq$ understandingOf.

Equivalently:

(1) *creation* $\subseteq$ *understanding,*
(2) *justification* $\subseteq$ *understanding,*
(3) *description* $\subseteq$ *understanding,* and
(4) *identification* $\subseteq$ *understanding.*

It shall be claimed that the role 'understandingOf', conceptually and logically, supports 'the [concept] *understanding* function based on the analysed [concept] understanding model (or *UND*)'. Similarly, we can define *CRN, JSN, DSN,* and *IDN* as sub-models of *UND* for representing *creation, justification, description,* and *identification,* respectively. Any of these models can, semantically, satisfy the terminologies and world descriptions in Table 4. Accordingly, relying on inductive rules, they can satisfy concept descriptions in Table 5.

Note that *CRN* (as a model) fulfils the desires of *UND* better (and more satisfying) than *JSN, DSN,* and *IDN.* Considering D_U as the understanding domain, we have:

$$D_U^{UND} \subseteq D_U^{CRN} \subseteq D_U^{JSN} \subseteq D_U^{DSN} \subseteq D_U^{IDN}.$$

More specifically:

- D_U^{CRN} represents the model of *creation* over the understanding domain. It consists of concepts which are (or could be) 'created' by human beings. Formally: $C^{CRN} \in D_U^{CRN}$.
- D_U^{JSN} represents the model of *justification* over the understanding domain. It consists of concepts which are (or could be) 'justified' by human beings. Formally: $C^{JSN} \in D_U^{JSN}$.
- D_U^{DSN} represents the model of *description* over the understanding domain. It consists of concepts which are (or could be) 'described' by human beings. Formally: $C^{DSN} \in D_U^{DSN}$.
- D_U^{IDN} represents the model of *Identification* over the understanding domain. It consists of concepts which are (or could be) 'identified' by human beings. Formally: $C^{IDN} \in D_U^{IDN}$.

Proposition. The terminological axioms and the world descriptions (in Table 4) and inductive concept descriptions (in Table 5) are all valid and meaningful for *CRN, JSN, DSN,* and *IDN*. Therefore, inductive concept descriptions are also valid and meaningful over the concatenation of the *creation, justification, description,* and *identification* functions that have supported these terminological models.

Proposition. All semantic satisfactions based on *IDN* are already satisfied by *DSN, JSN,* and *CRN* over D_U^{DSN}, D_U^{JSN}, and D_U^{CRN}, respectively. Informally, if one is able to focus on describing, justifying, and creating based on his conceptions of a concept, so, he is already capable of identifying that concept. Furthermore, he might be able to identify something else (some other phenomenon) with regard to his conception of that concept.

Formal Analysis. The formal semantics of the composite function '*creation (justification (description (identification (C))))*'—that is the product of the chain of functional roles—supports the proposed semantic model on D_U^{UND}, which is the central domain of concept understanding (central part of the understanding domain). Considering all the roles relevant for the concept *C*, we have:

$$(\forall R_1.C)^{CRN} = \{a \in D_U^{CRN} | \forall b.(a,b) \in R_1^{CRN} \to b \in C^{CRN}\}.$$

Therefore:

$$(\forall R_2.C)^{JSN} = \{a \in D^{JSN} | \forall b.(a,b) \in R_2^{JSN} \to b \in C^{JSN}\}.$$

Therefore:

$$(\forall R_3.C)^{DSN} = \{a \in D_U^{DSN} | \forall b.(a,b) \in R_3^{DSN} \to b \in C^{DSN}\}.$$

Therefore:

$$(\forall R_4.C)^{IDN} = \{a \in D_U^{IDN} | \forall b.(a,b) \in R_4^{IDN} \to b \in C^{IDN}\}.$$

In this formalism, R_1, R_2, R_3, and R_4 stand for creatingOf, justifyingOf, describ-ingOf, and identifyingOf, respectively. Consequently, CRN, JSN, DSN, and IDN have been interpreted roles of human beings. Accordingly, it's possible to represent the chain of functional roles in the form of the collection of the following implications:

$$(\forall R_1.C)^{CRN} \Rightarrow$$
$$(\forall R_2.C)^{JSN} \Rightarrow$$
$$(\forall R_3.C)^{DSN} \Rightarrow$$
$$(\forall R_4.C)^{IDN}.$$

It must be concluded that 'any role based on a conception of the concept C' to the left of any of implications ($\Rightarrow$) makes a logical premise for 'other roles based on the conceptions of the concept C' to the right of that implication. It shall be stressed that this is a very important terminological fact in semantic analysis of concept under-standing. The deduced logical relationship represents a stream of concept under-standing from deeper layers to shallower layers.

7   An Ontology for Concept Understanding

From the philosophical point of view, an ontology is described as studying the science of being and existence, see [13, 23]. Ontologies must be capable of demonstrating the structure of the reality of a thing/phenomenon. They check multiple attributes, par-ticularities, and properties that belong to a thing/phenomenon because of its structural existence. From another perspective and through the lenses of information and com-puter sciences, an ontology is an explicit [and formal] specification of a shared conceptualisation.

However, in my opinion, there are some conceptual relationships between these two descriptions of ontologies. Actually, ontologies in information sciences attempt to mirror the phenomena's structures in virtual and artificial systems. In fact, they focus on conceptual descriptions of phenomena's structures in order to provide proper backgrounds for specifications of their conceptualisations. Hence, the ontological descriptions in information sciences (and in knowledge-based systems) tackle to pro-vide appropriate logical and formal descriptions of a phenomenon's structure as well as its dependency to the other phenomena and to the environment. From this perspective, an ontology can be schemed and demonstrated by semantic networks and semantic representations. A semantic network is a graph whose nodes represent entities and whose arcs represent relationships between those entities.

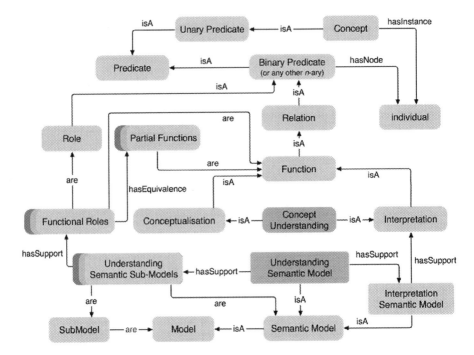

Fig. 1. An ontology for concept understanding.

According to [4], Fig. 1 represents a semantic network as an ontology for the phenomenon of 'concept understanding'. This hierarchical semantic representation:

- specifies the conceptual relationships between the most important ingredients of this research,
- demonstrates the logical representation of the phenomenon of 'concept understanding', and
- shows how the proposed semantic model attempts to represent concept understanding.

This semantic representation can be interpreted a specification of the shared conceptualisation of 'concept understanding' within terminological systems.

Note that the constructed ontology can be reformulated and formalised in ALC in the form of a collection of fundamental terminologies as following:

{
UnaryPredicate $\sqsubseteq$ Predicate,
BinaryPredicate $\sqsubseteq$ Predicate,
Concept $\sqsubseteq$ UnaryPredicate,
Concept $\sqsubseteq$ $\exists$hasInstance.Individual,
BinaryPredicate $\sqsubseteq$ ($\exists$hasNode.Individual $\sqcap$ $\exists$hasNode.Individual),
Role $\sqsubseteq$ BinaryPredicate,
Relation $\sqsubseteq$ BinaryPredicate,

Function ⊑ Relation,
Interpretation ⊑ Function,
Conceptualisation ⊑ Function,
ConceptUnderstanding ⊑ Interpretation,
ConceptUnderstanding ⊑ Conceptualisation,
PartialFunction ⊑ Function,
FunctionalRole ⊑ Role,
FunctionalRole ⊑ hasEquivalence.PartialFunction,
FunctionalRole ⊑ Function,
SubModel ⊑ Model,
SemanticModel ⊑ Model,
InterpretationSemanticModel ⊑ SemanticModel,
UnderstandingSemanticModel ⊑ SemanticModel,
UnderstandingSemanticSubModel ⊑ SubModel,
UnderstandingSemanticSubModel ⊑ SemanticModel,
InterpretationSemanticModel ⊑ ∃hasSupport.Interpretation,
UnderstandingSemanticModel ⊑ ∃hasSupport.InterpretationSemanticModel,
UnderstandingSemanticModel ⊑ ∃hasSupport.UnderstandingSemanticSubModel,
UnderstandingSemanticSubModel ⊑ ∃hasSupport.FunctionalRole
}

8 Concluding Remarks

Description Logics (DLs) attempt to provide descriptive knowledge representation formalisms to establish common grounds and interrelationships between human beings and machines. DLs have assisted me in revealing some hidden conceptual and logical assumptions about the phenomena of 'concept' and 'concept understanding'. More specifically, these assumptions can produce a better conceptualisation (and respectively, understanding) of 'concept understanding'. In this article, DLs have—by considering concepts as unary predicates and by applying terminological interpretations over them—proposed a realisable, as well as assessable, logical description for explaining the humans' concept understanding. Relying on such a logical description a theoretical model for concept understanding has been offered. The proposed model attempts to reflect the phenomenon of 'concept understanding' in terminological knowledge representation systems. It shall be concluded that the most significant contribution of the article has been providing a formal semantics for logical analysis of concept understanding. According to the logical analysis, a logical background for terminological representation of concept understanding has been expressed. Consequently, an ontology for 'concept understanding' has been designed and formalised. The offered ontology specifies my conceptualisation of the phenomenon of 'concept understanding'.

References

1. Baader, F., Calvanese, D., McGuinness, D., Nardi, D., Patel-Schneider, P.: The Description Logic Handbook: Theory, Implementation and Applications. Cambridge University Press, New York (2010)
2. Badie, F.: Concept representation analysis in the context of human-machine interactions. In: Proceedings of the 14th International Conference on e-Society. International Association for Development of the Information Society, Portugal (2016a)
3. Badie, F.: Towards concept understanding relying on conceptualisation in constructivist learning. In: Proceedings of the 13th International Conference on Cognition and Exploratory Learning in Digital Age. International Association for Development of the Information Society, Germany (2016b)
4. Badie, F.: A formal semantics for concept understanding relying on description logics. In: Proceedings of the 9th International Conference on Agents and Artificial Intelligence. SCITEPRESS Digital Library, Portugal (2017)
5. Barsalou, L.W.: Perceptual Symbol Systems. The Behavioural and Brain Sciences. Cambridge University Press, New York (1999)
6. Biggs, J.B., Collis, K.F.: Evaluating the Quality of Learning: The SOLO Taxonomy (Structure of the Observed Learning Outcome). Academic Press, New York (2014)
7. Blackburn, S.: The Oxford Dictionary of Philosophy. Oxford University Press, Oxford (2016). Web
8. Chaitin, G.J.: Algorithmic Information Theory. Cambridge University Press, New York (1987)
9. Davies, J., Fensel, D., van Harmelen, F.: Towards the Semantic Web, Ontology-Driven Knowledge Management. Wiley Online Publications, New York (2003)
10. di Pellegrino, G., Fadiga, L., Fogassi, L., Gallese, V., Rizzolatti, G.: Understanding motor events: a neurophysiological study. Exp. Brain Res. **91**, 176–180 (1992)
11. von Foerster, H.: Understanding Understanding, Essays on Cybernetics and Cognition. Springer, New York (2003)
12. Gray, P.M.D., Kulkarni, K.G., Paton, N.W.: Object-Oriented Databases - A Semantic Data Model Approach. Prentice Hall International Series in Computer Science. Prentice Hall, London (1992)
13. Stephan, G., Pascal, H., Andreas, A.: Knowledge representation and ontologies. In: Studer, R., Grimm, S., Abecker, A. (eds.) Semantic Web Services, pp. 51–105. Springer, Heidelberg (2007). https://doi.org/10.1007/3-540-70894-4_3
14. Honderich, T.: The Oxford Companion to Philosophy. Oxford University Press, Oxford (2005)
15. Jackendoff, R.: Semantic Structures. MIT Press, Cambridge (1990)
16. Kant, I.: Kritik der reinen Vernunft. VMA-Verlag, Wiesbaden. (imprint of the 1924 edt.), p. 967 et passim (1781)
17. Kintsch, W., Welsch, D., Schmalhofer, F., Zimny, S.: Sentence memory: a theoretical analysis. J. Mem. Lang. **29**, 133–159 (1990). Elsevier
18. MacKay, D.: Information Theory, Inference and Learning Algorithms. Cambridge University Press, New York (2003)
19. Peschl, M.F., Riegler, A.: Does representation need reality? Rethinking Epistemological issues in the light of recent developments and concepts in cognitive science. In: Riegler, A., Peschl, M., von Stein, A. (eds.) Understanding Representation in the Cognitive Sciences, pp. 9–17. Springer, Boston (1999). https://doi.org/10.1007/978-0-585-29605-0_1

20. Resnik, P.: Semantic similarity in a taxonomy: an information-based measure and its application to problems of ambiguity in natural language. J. Artif. Intell. Res. **11**, 95–130 (1999)
21. Schmidt-Schauss, M., Smolka, G.: Attributive concept descriptions with complements. Artif. Intell. **48**(1), 1–26 (1991). Elsevier
22. Simpson, J.A., Weiner, E.S.C.: The Oxford English Dictionary. Oxford University Press, Oxford (1989)
23. Staab, S., Studer, R.: Handbook on Ontologies, 2nd edn. Springer, Heidelberg (2009)
24. Uithol, S., van Rooij, I., Bekkering, H., Haselager, P.: Understanding motor resonance. J. Soc. Neurosci. **6**(4), 388–397 (2011). Routledge
25. Uithol, S., Paulus, M.: What do infants understand of others' action? A theoretical account of early social cognition. Psychol. Res. **78**(5), 609–622 (2014)
26. Webb, J.: Understanding Representation. Sage Publications, London (2009)
27. Zwaan, R.A., Taylor, L.J.: Seeing, acting, understanding: motor resonance in language comprehension. J. Exp. Psychol. Gen. **135**(1), 1–11 (2006)

Reasoning for Autonomous Agents in Dynamic Domains: Towards Automatic Satisfaction of the Module Property

Stephan Opfer$^{(\boxtimes)}$, Stefan Jakob$^{(\boxtimes)}$, and Kurt Geihs$^{(\boxtimes)}$

Distributed Systems Research Group, University of Kassel,
Wilhelmshöher Allee 73, Kassel, Germany
{opfer,s.jakob,geihs}@uni-kassel.de

Abstract. State-of-the-art service robots that fetch a cup of coffee and clean up rooms require cognitive skills such as learning, planning, and reasoning. Especially reasoning in dynamic and human populated environments demands for novel approaches that can handle comprehensive and fluent knowledge bases. Our long-term objective is an autonomous robotic team that is capable of handling dynamic and domestic environments. Therefore, we combined ALICA – A Language for Interactive Cooperative Agents – with the Answer Set Programming solver Clingo. The answer set programming approach offers multi-shot solving techniques and non-monotonic stable model semantics, but requires to keep the Module Property satisfied. We developed an automatic satisfaction of the Module Property and chose topological path planning as our evaluation scenario. We utilised the Region Connection Calculus as the underlying formalism of our evaluation and investigated the scalability of our implementation. The results show that our approach handles dynamic environments and scales up to appropriately large problem sizes while automatically satisfying the Module Property.

Keywords: Answer Set Programming · Region Connection Calculus
Module Property · Multi-shot solving

1 Introduction

Due to the development of autonomous vacuum cleaners, lawnmowers and pool cleaners, autonomous robots stepped into our everyday life and also in other areas similar developments are currently taking place. Automated guided vehicles that take care of the logistics in production plants or parcel service centres are already commonly used [1], but also most car manufacturers are developing autonomous cars [2]. Even autonomous and interactive toys become more and more intelligent and conquer our children's rooms [3,4].

In contrast to these single purpose devices, researchers in the field of service robots focus on multi-purpose robots. Figure 1 shows a state-of-the-art service robot, which can do everyday household tasks. The variety of those tasks pose

© Springer International Publishing AG, part of Springer Nature 2018
J. van den Herik et al. (Eds.): ICAART 2017, LNAI 10839, pp. 22–47, 2018.
https://doi.org/10.1007/978-3-319-93581-2_2

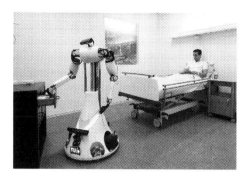

Fig. 1. Domestic service robot [5].

new challenges to the research community because the number of elements relevant to the robot's environmental representation is tremendous. Further challenges arise from human populated environments, that require robots to handle very dynamic situations, because human beings insert, remove, and displace objects in their environment. Furthermore, human beings are themselves dynamic obstacles from the robot's point of view. In order to cope with such environments, robots need cognitive capabilities such as learning, planning, and reasoning. Reasoning about such complex and dynamic domains, e.g., requires a suitable level of abstraction in order to make the reasoning more tractable. Consider the example of a robot that should fetch a cup. Reasoning about possible positions of the cup and how the robot could get there, should not be polluted by the cup's exact coordinates or the robots trajectory planning. The robot should only take its current position in the building, locked doors, obstacles, and the building's topology into account.

Symbolic knowledge representation and reasoning is a common approach for tackling such problems [6]. Nevertheless, most benchmarks present today[1] are designed in a way that prevents the development of solvers that continuously use a changing amount of knowledge while solving different problem instances over time. Our contribution is to enable multi-agent systems to continuously reason about dynamic environments by utilising Answer Set Programming (ASP) – a non-monotonic knowledge representation and reasoning formalism [7], suitable for multi-shot solving [8]. In our case, the multi-agent system is controlled by the ALICA – A Language for Interactive Cooperative Agents [9]. Our preliminary work [10] forms the basis for integrating ASP with ALICA and allows us to extend it with a general solver interface which in turn makes ALICA open for a wider set of application domains. Moreover, we had to extend the utilised ASP solver with query mechanism according to [11]. In contrast to our preliminary work [12], the Module Property is now satisfied automatically. In order to evaluate our query mechanism, we used an ASP-based path planning scenario that utilizes a simplified version of the Region Connection Calculus [13], denoted as RCC-4.

---

[1] http://www.satcompetition.org/ [Online; accessed September 12, 2017].
http://aspcomp2015.dibris.unige.it/ [Online; accessed September 12, 2017].

The remainder of this paper is structured as follows. Section 2 introduces ALICA, ASP, and RCC-4. The integration of ALICA with ASP is described in Sect. 3. Furthermore, the query semantics extension including the automatic satisfaction of the Module Property is elaborated in Sect. 4. Section 5 provides the description of our evaluation scenario, whose results are presented in Sect. 6. Finally, we compare our work with other approaches in Sect. 7 and conclude with Sect. 8.

2   Foundations

This section is divided into three subsections. In Sect. 2.1 the focus is set on concepts of ALICA that are necessary to understand in the context of this work. The same holds for Sect. 2.2 that is about the syntax and semantics of ASP. In Sect. 2.3 an explanation of the basic relations of the Region Connection Calculus 4 is given.

2.1   ALICA

The ALICA framework is designed to coordinate a cooperative team of autonomous agents. Explaining all features of this framework is beyond the scope of this work and we would like to point the interested reader to the dissertation of Skubch [9] and two supplementary publications [14,15]. In this section, our goal is to explain the fundamental principles of ALICA and focus on the parts that we changed to make a wider set of general problem solvers accessible from within an ALICA program.

The ALICA framework is distributed in the sense that each agent in the team is running its own independent ALICA behaviour engine. Each behaviour engine determines the sequence of actions of the local agent while coordinating itself with other engines and taking the current situation as well as a given ALICA program into account. Sometimes frameworks like ALICA are also denoted as sequencers [16].

An ALICA program is a special tree, whose interior nodes are plans and its leaf nodes are atomic behaviours. The *CleanUp* plan in Fig. 2 is an example of such an interior node. A plan can include several states $(Z_0 \ldots Z_8)$ that are connected with guarded transitions to create finite state machines (FSM). Each FSM is annotated with a task (*Tidy Up, Wipe Floor, Inspect*) and a pair of cardinalities for the minimum and the maximum number of agents allowed in the corresponding FSM. Each state of a plan, except for terminal states (Z_4), can contain behaviours and plans that represent leaf or interior nodes respectively on the next level of the tree. In Fig. 2 plans and behaviours are distinguished by the colour of their boxes, e.g., state Z_0 contains the plan *Drive*, which is blue, and state Z_1 contains the behaviour *Pick Up*, which is orange. It is important to note that a plan, which is referenced in a state, is a complete plan like the *Clean Up* plan itself and therefore can include state machines with other behaviours and plans.

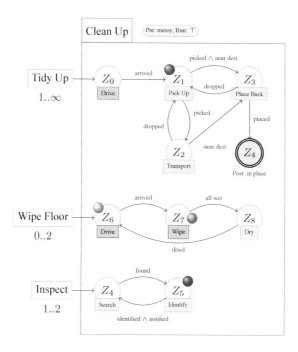

Fig. 2. Simple clean up plan [12]. (Color figure online)

The coloured circles on top of some states in Fig. 2 illustrate a possible global execution state of the plan. Each circle represents an agent. The red circle, for example, could be the local agent executing the *Pick Up* behaviour, while the other circles represent other agents in the team, whose corresponding behaviour engine have published their execution state to the local agent.

In order to understand the extension of ALICA by a general problem solver interface, it is necessary to introduce the notion of ALICA plan variables. For simplicity let us modify the *Inspect* state machine from Fig. 2 to create a plan on its own.

The purpose of this plan, as shown in Fig. 3, could be the identification of coffee cups that should be cleaned, because they are dirty and not used any more. Therefore, an agent searches for coffee cups in state Z_4 and switches to state Z_5, in order to classify them. For remembering which cup was found in state Z_4 the plan variable X can be set to the corresponding cup and thereby influence

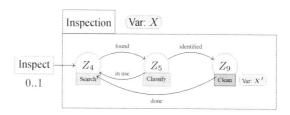

Fig. 3. Inspection plan [12].

the agent's behaviour during the rest of the plan execution. This influence can even reach to deeper levels of the plan tree, as indicated by variable X' of the *Clean* plan in state Z_9. It is possible to define variable bindings over states in an ALICA plan tree, e.g., stating that X denotes the same variable as X'. This allows determining the agent's behaviour in the *Clean* plan, depending on which cup was found in the *Inspection* plan.

ALICA, as presented in [9], only provided one solver for assigning values to plan variables. The given solver addresses non-linear continuous constraint satisfaction problems and, as it was hard-wired to ALICA, the applicability of ALICA was limited for some domains. Our extension of ALICA with a general solver interface tackles this issue (see Sect. 3).

2.2   Answer Set Programming

Answer Set Programming (ASP) is a declarative approach for solving NP-search problems. It can be seen as the result of decades of research in the areas of knowledge representation, logic programming, and constraint satisfaction. Thereby, its focus is on expressiveness, ease of use, and computational effectiveness [17]. An ASP program is specified by a set of rules of the form $a_0 \leftarrow a_1 \wedge \ldots \wedge a_m \wedge \sim a_{m+1} \wedge \ldots \wedge \sim a_n$. Each a_i denotes a predicate $p(t_1, \ldots, t_k)$ with terms $t_1, \ldots, t_k$ build from constants, variables, and functions. Rules consist of three parts, namely the head a_0, the positive part $a_1 \wedge \ldots \wedge a_m$ and the negative part $\sim a_{m+1} \wedge \ldots \wedge \sim a_n$ of the body. The semantics of the default negation $\sim$ is that of negation-as-failure. That means, $\sim a_i$ is considered to hold if it fails to prove that a_i holds. Nevertheless, ASP also provides classic negation $\neg a_i$, whose semantics is that $\neg a_i$ holds, if a_i does not.

The example in Listing 1.1 is a syntactically correct ASP program. In order to create syntactically correct ASP programs, the rules have to be transformed in the following way: $\leftarrow$ is transformed to `:-`, $\wedge$ are replaced by `,` and a rule is ended by a dot. The `-` in front of `robot(X)` stands for classic negation $\neg$ and `not` in front to of `broken(X)` means default negation $\sim$. The `;` is a syntactic shortcut for creating several rules at once. Rule 1, therefore, creates `robot(chuck)`, `robot(fox)`, and `robot(lisa)`. Furthermore, Rules 1 and 2 have an empty body. So their heads are unconditionally true, and they are denoted as facts. Rule 3 makes use of the default negation $\sim$ and states that a robot can drive, as long as it cannot be proven, that it is broken (`not broken(X)`). Here X is a variable (starts with a capital letter), which can be substituted with any element of the Herbrand universe of the given program.

```
1    robot(chuck; fox; lisa).
2    broken(fox).
3    canDrive(X)  :-  robot(X),not broken(X).
4    highFailureRate  :-  Working = #count{X : canDrive(X)},
     Broken = #count{X : broken(X)}, Working < Broken.
```

Listing 1.1. Robots can drive, as long as they are not broken.

The Herbrand universe of the program in Listing 1.1 only contains four constants (start with lower case letter): {chuck, fox, lisa, highFailureRate}. Rule 4 derives the constant highFailureRate, if there are more broken robots than driving ones. As shown by this rule, ASP is capable of handling integers and provides aggregate functions like #count or #sum and arithmetic functions like < or +. The result of this program will state, that there are three robots of whom chuck and lisa can drive, the constant highFailureRate does not occur.

State-of-the-art ASP solvers [7,18] usually work in two steps. First, they ground the program and afterwards determine all stable models of the grounded program. A program, as well as every part of it, is grounded if it does not contain any variable. In order to create a grounded program, informally speaking, the variables of each rule are replaced by each possible substitution with an element of the program's Herbrand universe. The Herbrand universe of a program is constructed from all constants and functions occurring in the program. Grounding a program that way would increase the number of rules enormously, therefore the utilised grounding algorithms try to keep the grounded program as small as possible, without altering the programs meaning. For example, Rule 3 of Listing 1.1 will not be part of the grounded program, if there is no robot available.

Solving a grounded program is often done with SAT solving techniques that are adapted to the stable model semantics of ASP. A model in ASP is a set M of ground predicates that for every rule either contains the rule's head ($a_0 \in M$), or does not include all predicates of the positive part of the rule's body ($\{a_1, \ldots, a_m\} \not\subseteq M$), or contains predicates from the negative part of the rule's body ($\{a_{m+1}, \ldots, a_n\} \cap M \neq \emptyset$). Informally speaking, a stable model is as small as possible and contains predicates, only if they are justified by facts. For a detailed introduction to the stable model semantics, see [19].

In our approach, we choose the Clingo 4.5.3 ASP solver [7], which introduces the notion of *External Statements* to ASP [8,20]. *External Statements* in combination with *Program Sections*, explained later on, are the key concepts for enabling the query semantics described in Sect. 4. The *External Statements* are predicates annotated with #external (see Listing 1.2). Those predicates are not removed from the body of a rule during grounding, even if they do not appear in the head of any rule, because of this annotation. Furthermore, it is possible to set them explicitly to true or false without an extra grounding step.

```
1    #external closed(lab,hall).
2    connected(lab,hall) :- not closed(lab,hall).
3    disconnected(lab,hall) :- closed(lab,hall).
```

Listing 1.2. Modeling a door using an External Statement.

The example, given in Listing 1.2, is part of our evaluation scenario presented in Sect. 5 and will also be used in ASPExtensionQueries presented in Sect. 4.1. In this example closed(lab, hall) is marked as an *External Statement* and is therefore set to false by default. If the *External Statement* closed(lab, hall) is set to true, the head of Rule 3 holds and the predicate disconnected(lab, hall) is part of the stable model. The head of Rule 2 cannot be derived since

closed(lab, hall) is set to be true. Usually, during the grounding procedure
Rule 2 would be removed, because its body cannot be derived. However, since
this rule contains an *External Statement*, it stays part of the grounded logic program. Therefore, it is possible to change a logic program without another grounding step by using *External Statements*. For example, if the *External Statement*
closed(laboratory, hall) is set to true, the laboratory and the hall are disconnected from each other. If it is set to false the predicate disconnected(lab,
hall) no longer holds but connected(lab, hall) can be derived, thus the
rooms are connected. Furthermore, changing the *External Statement* does not
change the size of the resulting stable model.

Additionally, Clingo introduces *Program Sections* [7]. *Program Sections* are
used to divide a logic program into different parts, which can be grounded separately. An example is given in Listing 1.3.

```
1    #program rcc4_composition_table.
2    disconnected(X,Z) :- properPart(X,Y),disconnected(Y, Z),
     X != Z.
3    #program rcc4_facts.
4    properPart(office1,offices).
5    disconnected(offices,studentArea).
```

Listing 1.3. Usage of program sections.

This example contains two *Program Sections* identified by the #program
prefix, i.d., rcc4_composition_table and rcc4_facts. Moreover, the order in
which they are grounded influences the facts, which appear in the stable models. If the *Program Section* rcc4_composition_table is grounded before the
Program Section rcc4_facts, the resulting model would only contain the facts
properPart(office1,offices) and disconnected(studentArea, offices).
If *Program Section* rcc4_facts is grounded first, the model will contain these
facts before the *Program Section* rcc4_composition_table is grounded. Once
this *Program Section* is grounded the model will additionally contain the fact
disconnected(office1, studentArea), since this rule's body holds for the
grounding of X by office1, Y by offices, and Z by studentArea.

2.3   Region Connection Calculus

In this section, the base relations of the Region Connection Calculus 4 (RCC-
4) are shown, which is based on the Region Connection Calculus 8 presented
by Randell et al. in [13]. We have been inspired to use RCC-4 instead of using
RCC-8 by the implementation of RCC-4, that is provided by the QSRlib Foundation [21] since the majority of the RCC-8 relations were not used for modelling
the Distributed Systems Department (see Sect. 6). These calculi are commonly
used in qualitative spatial reasoning and will be used to model our evaluation
scenarios in Sect. 5. The foundation of the relations is the binary relation $C(x,y)$,
which expresses that two spatial regions of unknown size are connected. Informally speaking, they share at least one common point. Furthermore, $C(x,y)$ is

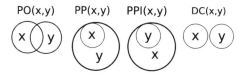

Fig. 4. RCC-4 base relations [12].

reflexive and symmetric. By using the C(x,y) relation four base relations are defined and shown in Fig. 4.

Two regions are partially overlapping (PO), if they are connected (C), meaning they share a common point, region or part of their border. Region x is a proper part (PP) of Region y if y contains Region x, which means that Region x is connected to y and no part of x is outside the border of y. Since this relation is not symmetric the inverse relation PPI is included as well. Additionally, two regions are disconnected (DC) if they do not share a common point. By using the composition table shown in Table 1 the transitive relations between Region x and z can be derived, given the relations between the pairs (x,y) and (y,z). Hereby, * denotes that all four relations can hold. For example, if Region x is a proper part (PP) of Region y and Region y is disconnected (DC) from Region z, it can be derived that Region x is disconnected (DC) from Region z.

Table 1. RCC-4 composition table [12].

	DC	PO	PP	PPI
DC	*	DC PO PP	DC PO PP	DC
PO	DC PO PPI	*	PO PP	DC PO PPI
PP	DC	DC PO PP	PP	*
PPI	DC PO PPI	PO PPI	PO PP PPI	PPI

An example for using RCC-4 to model a building is given in Sect. 5. Furthermore, the RCC-4 relations can be used to model the relation of objects to each other. For example, in the domain of domestic service robots RCC-4 can be used to model objects on a table without defining their exact positions, e.g., a cup and a plate could be proper parts of the table and could be disconnected from each other if they are not touching each other.

3   Extending ALICA with a General Solver Interface

In this section, we describe the extension of the ALICA framework with a general solver interface, in order to integrate different solvers into the ALICA framework. By now, ALICA has only been able to use a gradient solver, which is able to solve non-linear continuous constraint satisfaction problems.

We created an abstract solver interface that allows solving very different problem classes. A solver that is compatible with this interface, has to adhere to four corresponding concepts: *solver*, *variable*, *term*, and *problem descriptor*. The *solver* is expected to solve the problem contained in a *problem descriptor*. A *problem descriptor* encapsulates the other two concepts that describe the actual problem. Solving a problem means to assign values to *variables* in a way that the values fulfil constraints described by a set of *terms*. The ALICA engine only understands these basic relations between the four concepts. The type of values that can be assigned to a specific *variable* and the format of the *terms* is unknown to the ALICA engine. Only the domain specific parts of an ALICA program should be able to understand the actual meaning of a described problem and be able to interpret the solver's results. Furthermore, the *variable* concept is identical to that of ALICA plan variables (see Sect. 2.1). Therefore, the ALICA engine is able to collect all relevant constraints (*terms*) over *variables* in the plan hierarchy, when a solution for a specific *variable* is required and can pass it via a *problem descriptor* to the responsible *solver*.

There are two methods for interacting with the solver interface. The first method checks if a solution for a given set of *variables* can be found. This method only determines, whether a solution exists and does not provide the solution itself. Usually, the complexity of providing a solution is the same as for proofing only the existence of a solution, but in case of an optimisation problem, it is not the same. Determining, for example, whether there is a cup available, is much easier than to determine the closest available cup. Therefore, whenever it is enough to check the existence of a solution the ALICA engine can save some time doing so.

With the second method, it is possible to actually determine the solution itself. The solver assigns its result to the corresponding variables and the ALICA engine hands them, via the *problem descriptor*, back to the calling domain-specific part of its program. The gradient solver, for example, returns continuous values and an ASP solver could return a set of predicates.

So far, the presented interface has been used to integrate three different solvers. The first solver is the gradient solver, which was a part of the original ALICA framework and has been adapted to the new interface. Additionally, an ASP solver, which is presented in this work, is integrated by using this interface. Furthermore, Witsch presents in [22] a middleware that enables a decision-making process for a group of robots. This middleware uses the presented interface to exchange variables and proposals between agents.

4   Extending Clingo with Query Semantics

The extension of Clingo with query semantics is twofold. On the one hand, there is the structure of the query itself and on the other hand there is the processing of such a query. The query structure holds all information necessary for processing the query, e.g., ASP rules that change the ASP program during the query process. The details are explained in Subsect. 4.1. The workflow during the processing of a query is described in Subsect. 4.2. This subsection places the focus especially on the requirements and techniques for changing the ASP program only during the query process and how it is possible to reestablish the programs state without the changes inflicted due to the query.

4.1 Query Structure

The central part of the query structure, further denoted as ASPQuery, is the ASPTerm. In compliance with the *Term* and *Variable* concepts, described in Sect. 4, the ASPTerm constraints a set of ASPVariables with a set of user defined rules. The interpretation of theses rules depends on the type of the ASPQuery. It is either an ASPFilterQuery or an ASPExtensionQuery (see Fig. 5).

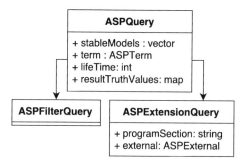

Fig. 5. Classmodel of ASPQueries.

The ASPQuery does, apart from the ASPTerm, also include a lifetime, a list of stable models, and a map of results with its corresponding truth values. The lifetime determines the number of solving operations the query should be processed with the current ASP program. The list of stable models includes all stable models from the last solving operation. Finally, the map of results and truth values contains one entry per queried predicate. The queried predicate is mapped to `true` if it is part of all stable models, `false` if its classic negation is part of all stable models, and `unknown` in all other cases.

The ASPFilterQuery is a simple ASPQuery that only filters the stable models by a set of given predicates. In the opposite to the ASPExtensionQuery, the ASP program is not altered due to ASPFilterQueries. Gelfond et al. [11] present a similar query mechanism, that checks whether a grounded predicate is part of the solver's stable models. In extension to this query formalism, we allow wildcards as parts of the filters. The filter `robot(wildcard)`, e.g., filters for all grounded `robot/1` predicates, including classic negated predicates. The truth value of this filter is set to `true` if at least one grounded `robot/1` predicate exists in all stable models, `unknown` if no grounded `robot/1` predicate appears in any stable model, and `false` otherwise.

The ASPExtensionQuery is a much more sophisticated way to query an ASP program. During it lifetime, it changes the ASP program and therefore its stable models. Afterwards, the ASP program is reverted to its original state. Consider, for example, an ASP program that identifies cups as free to use, when they are on the cupboard. With the ASPExtensionQuery, it is possible to temporarily declare all cups in the dishwasher as free to use, too. In general this is done by adding an arbitrary set of rules, given by the ASPTerm, to the ASP program. One rule in the set is specially handled and therefore further denoted as query rule. The head of the query rule defines, similar to the filter of the ASPFilterQuery, the crucial predicates to look for in the stable models.

For each ASPExtensionQuery a unique *External Statement* and *Program Section* is created. Both are necessary to revert the ASP program back to its original state after the end of the query's lifetime. The *External Statement* is added to the body of each rule and all rules are added to the ASP program as part of the *Program Section*. The *Program Section* is grounded one time before the solving operations and as soon as the lifetime of the query is expired, the *External Statement* is released. Releasing the *External Statement* automatically removes all added rules from the ASP program and therefore their influence on the stable models.

Consider the following example, given in Listing 1.4. It demonstrates the setup of an ASPExtensionQuery in pseudo code.

```
1 term->addQueryRule("goalReachable(X):-reachable(X,Y),
        goal(X),start(Y).")
2 term->addRule("goal(r1405B).")
3 term->addRule("start(r1411).")
4 query->setTerm(term)
5 query->programmSection("distributedSystems")
6 query->external(distributedSystemsExt)
7 query->lifeTime(1)
```

Listing 1.4. Example of an ASPTerm.

The example is part of an ASP navigation, which will be explained in detail in Sect. 5. Here it is queried whether room r1405B can be reached from room r1411. Therefore, the query rule and two facts are added to the *Program Section* distributed Systems and guarded by the *External Statement* distributedSystemsExt. Finally, the query's lifetime is set to one, which means that the query will be removed from the ASP program directly after it has been answered one time.

Since the ASPExtensionQuery is used to alter an ASP program, it can violate the Module Property, whose satisfaction guarantees that two *Program Sections* can be combined without rendering the ASP program unsolvable. A Module $\mathcal{P}$ is defined as a triple of sets (P, I, O). P is a ground program over the universe ground(A) and both, I and O, are disjoint subsets of ground(A). Furthermore, all *atoms* appearing in P are either part of I or O and all rule heads are part of O. I and O are denoted as input and output, respectively. Given this definition of a Module, two Modules $\mathcal{P}$ and $\mathcal{Q}$ are compositional, meaning their join will not violate the Module Property if the following two conditions hold. The first condition is that the output sets of both Modules are disjoint, meaning that they do not share a common *predicate*. The second condition relies on strongly connected components [23]. A strongly connected component is a subset of a directed graph, in which every vertex is reachable by any other vertex in this subset. In order to check this condition, all strongly connected components of the union of $\mathcal{P}$ and $\mathcal{Q}$ (SCC) have to be considered. If any strongly connected component in SCC has a non-empty intersection with both output sets ($O(\mathcal{P})$ ∩ SCC ≠ ∅ or $O(\mathcal{Q})$ ∩ SCC ≠ ∅), this condition is violated. Violating any of these two conditions would result in a recursion between both Modules and therefore violating the Module Property. The given definition is based on the lecture material from Schaub [24].

Expanding an ASP program with a *Program Section* can violate of the Module Property, which causes that the ASP program can no longer be grounded and solved. Therefore, the *ASPExtensionQuery* has to guarantee unique rule heads for every rule in the query. This can be done by enclosing the rule heads inside a unique predicate, which in this case is realised by a constant string and a counter provided by the *ASPSolverWrapper* (see Sect. 4.2). This way of satisfying the Module Property does not change the arity of the query rule head but replaces it with a new predicate, which has to be considered when the result is returned to the user. In order to cope with this problem and making the Module Property transparent for the user, the *ASPExtensionQuery* is expanded by an automatic satisfaction of the Module Property. The pseudo code in Algorithm 1 describes our approach to guarantee the satisfaction of the Module Property.

This algorithm uses an *ASPExtensionQuery* q and the value of the counter c, which is maintained by the *ASPSolverWrapper*, as inputs and returns a modified unique *ASPExtensionQuery*. The first two steps in this algorithm create a unique *Program Section* ps and a unique *External Statement* ex, which are used in the following steps. In Step 3 (Lines 3–5), the query's facts appearing in the query rule's bodies are encapsulated in a new predicate pred, which is a combination of ps and c. After this step is completed, the query rule is expanded by the unique *External Statement*, which allows the removal of the query after it has been answered. Step 4 duplicates the query rule q', which will be explained based on an example in the next paragraph. Furthermore, in Step 5 (Lines 8–10) the occurrences of all rule heads are encapsulated in the predicate pred, as well, which marks the end of altering the query rule. The additional rules and facts have to be adapted since they can still violate the Module Property. Therefore, Step 6 (Lines 11–14) alters the rule heads by encapsulating them in pred and by adding ex to the body of every rule. This creates unique rules, which can later be removed from the ASP program by releasing the *External Statement*. The same process is used in Step 7 for altering the query's facts, which results in a unique query.

An example for the application of this algorithm is presented in Listing 1.5. This example is part of the evaluation scenario presented in Sect. 5. In this example, the counter c, managed by the *ASPSolverWrapper*, is set to 1.

Algorithm 1: Automatic satisfaction of the Module Property.

Input : ASPExtensionQuery q, Counter c
Output: ASPExtensionQuery with unique rules

1 Create unique Program Section ps
2 Create unique External Statement ex
3 **foreach** *fact in q.facts* **do**
4 | replace occurence of fact in q.queryRule.body by ps(fact)
5 **end**
6 add(q.queryRule.body, ex)
7 q' = duplicate q.queryRule
8 **foreach** *rule in q.rules* **do**
9 | replace occurence of rule.head in q' by ps(rule.head)
10 **end**
11 **foreach** *rule in q.rules* **do**
12 | replace rule.head by ps(rule.head)
13 | add(rule.body, ex)
14 **end**
15 **foreach** *fact in q.facts* **do**
16 | replace fact in q.queryRule by ps(fact)
17 | expand(fact, ex)
18 **end**
19 **return** *q*

```
 1 //ASPExtensionQuery before applying the algorithm
 2 goalReachable(X) :- reachable(X, Y),goal(X),start(Y),
       room(X),room(Y).
 3 reachable(r1405,r1406) :- room(r1405),room(r1406).
 4 goal(r1405B).
 5 start(r1411).
 6
 7 //ASPExtensionQuery after applying the algorithm
 8 #program query1.
 9 #external extQuery1.
10 query1(goalReachable(X)) :- reachable(X,Y),query1(goal(X)),
       query1(start(Y)),room(X),room(Y),extQuery1.
11 query1(goalReachable(X)) :- query1(reachable(X, Y)),
       query1(goal(X)),query1(start(Y)),room(X),room(Y),extQuery1
       .
12 query1(reachable(r1405,r1406)) :- room(r1405),room(r1406),
       extQuery1.
13 query1(goal(r1405B)) :- extQuery1.
14 query1(start(r1411)) :- extQuery1.
```

Listing 1.5. Application of the automatised satisfaction of the Module Property.

The example consists of the query rule in Line 1, an additional rule stating that if room 1405 and room 1406 exist they are reachable by each other, and two facts defining the start and goal position of the navigation. These facts can cause a violation of the Module Property since they form two Strongly Connected Components between two Modules if this query is used twice. Therefore, the automatized satisfaction of the Module property is used. The result of this process is shown in Lines 8 till 14 of Listing 1.5. Hereby, the Lines 8 and 9 have already been added in the previous version of the *ASPExtensionQuery* presented in [12], which is a unique *Program Section* and *External Statement*. This part of the query is followed by the query rule and its duplicate in Line 10 and 11. In both cases, the heads are expanded by a new predicate `query1` that is identical to the *Program Section*, rendering the heads unique. Furthermore, the appearance of every fact is replaced by its expanded version, which is shown in the Lines 13 and 14 of this example. The only difference between the duplicates is the handling of the rule heads of additional rules like Line 12. In order to satisfy the Module Property, the head of this rule has to be expanded as well. This change has to be conducted in the query rule, too, in order to use the corresponding predicate. This can cause problems when solving the query. In this example the `reachable` predicate is expanded by the rule `reachable(r1405, r1406):- room(r1405), room(r1406)`. Therefore, the occurrence of the rule head in the query rule has to be replaced. This causes that the `reachable` predicates in the knowledge base are no longer appearing in the query rule. Thus, leaving out possible solutions for the query. To cope with this problem, the duplicated query rule is used. The appearance of the additional rule heads is not replaced, leaving the possibility to use the knowledge base. After solving the query the unique predicate is removed from the results.

To sum up, the presented approach of an automatized satisfaction of the Module Property allows the user to create *ASPExtensionQueries*, without the opportunity to violate it. Furthermore, this approach allows to expand predicates appearing in the knowledge base without loosing the possibility to use the knowledge base. As a last point, this approach is transparent to the user, who formulates the query and is given the result without any additional predicates encapsulating the original query.

4.2 Workflow of Queries

In order to use the described ASPTerm and ASPQueries, the query structure has to be accessed by the ALICA behaviours and the ASPQueries have to be forwarded to the Clingo ASP solver. Therefore, a wrapper has been created, that encapsulates the Clingo ASP solver and provides access to the query structure. This wrapper, named *ASPSolverWrapper*, is registered at the ALICA framework by using the interface explained in Sect. 3 and can be used by ALICA behaviours. The interaction workflow between the ALICA Engine and the wrapper are depicted in Fig. 6.

As mentioned in Sect. 2.1, an ALICA program consists of a directed acyclic plan tree. The ASPTerms needed to formulate queries are created inside the runtime conditions of such plans. Once the ASPTerm is created (an example is given in Listing 1.4), it can be used by an ALICA behaviour to formulate a query. Thereby, the *External Statements'* truth values are given by a worldmodel class,

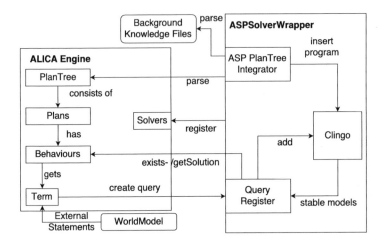

Fig. 6. Query workflow [12].

which encapsulates data that was perceived by the local agent or received from
other agents, e.g., the state of a door. After the ASPQuery has been formulated,
it is registered at the wrapper, which passes it to the ASP solver Clingo. In case
of an ASPExtensionQuery, the automatized satisfaction of the Module Property
(Sect. 4) is applied, which encapsulates the query in a unique *Program Section*.
Afterwards, the wrapper adds the query rule, the set of additional rules, and
facts to the solver's ASP program. Additionally, the ASPPlanTreeIntegrator
parses the ALICA program's plan tree to enable reasoning about its structure,
which is done by rules given in the background knowledge files. These files, for
example, contain rules to detect malformed ALICA programs. The corresponding
ASP rules are presented in [10]. The ASPPlanTreeIntegrator is only used during
the first registered query since the predicates stay part of the following stable
models as soon as they have been grounded. Furthermore, this increases the
runtime of the first query but reduces the runtime of the following queries, since
this *Program Section* has not to be grounded anymore. Once all program parts
(queries, background knowledge, and plan structure) have been added, Clingo
grounds and solves the program. The in this process derived stable models are
passed to the registered queries and saved to enable further use by other parts
of the ALICA framework, especially the ALICA behaviours. Finally, the results
are returned to the ALICA behaviour via two methods defined by the created
interface. The first method is named `existsSolution`, which checks the truth
value of the query without returning stable models or ground predicates. This
method can be used in combination with an ASPFilterQuery, which checks if
an ASP predicate is part of at least one or all stable models since the caller
is only interested if the queried fact is part of the stable models. The second
method is named `getSolution`. This method is used to return the derived stable
models to the querying ALICA behaviour. This method can, for example, be
used in combination with an ASPExtensionQuery, where a query rule, a set

of additional rules, and facts are used to alter the ASP program. Since the
ASPExtensionQuery modifies the ASP program the resulting stable models and
the queried rule heads are returned to the querying ALICA behaviour. Since the
rule heads have been altered by Algorithm 1, the encapsulating unique *predicate*
is removed before the results are returned to the caller. After the results are
returned, the ALICA behaviour can react to the changes in the model or to the
resulting rule heads. By returning the calculated results to the ALICA behaviour,
the workflow of a query is finished and the queries lifetime is reduced by one
and the ALICA behaviour can create the next query.

5   Evaluation Scenario

Our approach for handling dynamic domains will be evaluated using the scenario
presented in this section. The base of this scenario is a map of the Distributed
Systems Department of the University of Kassel and was created by using a
TurtleBot [25]. A TurtleBot is a small service robot equipped with a 3D camera
and was in our case extended with a 2D laser scanner. The customised version
of a TurtleBot is depicted in Fig. 7 alongside the resulting map, which is shown
in Fig. 8.

Fig. 7. Adapted version of a TurtleBot.

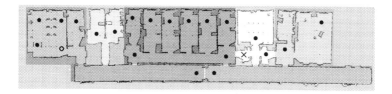

Fig. 8. Map of the distributed systems department [12]. (Color figure online)

This map consists of 19 rooms and is divided into seven `areas`, which are highlighted in different colours. These areas include from left to right `studentArea` (red), `mainHallA` (black), `workshop` (green), `offices` (blue), `mainHallB` (purple), `utility` (yellow) and `organization` (orange). Additionally, 56 points of interest (POI), examples marked with dots, are placed on this map. For example, a POI was placed on different workplaces, the coffee machine or the conference room. Furthermore, the robot's position is marked with a circle and the navigation goal used in this scenario is marked by a cross. The relations between the areas, rooms and POIs are modelled using the Region Connection Calculus 4 (Sect. 2). A POI is a `properPart` of a specific room and `disconnected` to all other rooms. Rooms are either `partialOverlapping` with other rooms, `properPart` of areas or `disconnected` from both. Areas can either be `partialOverlapping` or `disconnected`. Additionally, doors have been modelled utilising *External Statements*, as shown in Listing 1.2. This enables our approach to use different parts of the logic program without an additional grounding step and thus allows to change the stable models. This should decrease the runtime for answering ASPQueries and keep the models' size stable since no additional predicates have to be added to open or close doors, which could lower the performance of the ASP program. In order to use this scenario, we implemented an ALICA behaviour, which uses an ASPFilterQuery to check if the robot's goal position (cross) is reachable from its current position. In this case, we use the transitive closure defined by the predicate `reachable(X,Y)`, which is presented in Listing 1.6 as a simple path planning approach.

```
1  reachable(X,Y) :- reachable(Y,X),X!=Y.
2  reachable(X,Z) :- reachable(X,Y),reachable(Y,Z),
                     X!=Y,Y!=Z,X!=Z.
3  reachable(X,Y) :- partialOverlapping(X,Y),room(X),room(Y).
4  reachable(X,Y) :- partialOverlapping(X,Y),area(X),area(Y).
5  reachable(X,Y) :- properPart(X,Y),room(X),area(Y).
```

Listing 1.6. Transitive closure of reachable relation.

Rules 1 and 2 of the listing express, the `reachable` relation is symmetric and transitive. Furthermore, Rules 3 and 4 state, that a room `room(X)` or area `area(X)` is `reachable` from another room `room(Y)` or area `area(Y)` if they are partial overlapping `partialOverlapping(X,Y)`. As a last point, a room is `reachable` from an area if the room is a `properPart` of the area (Rule 5).

6 Evaluation

In this section, we present the revised and updated evaluation results from [12]. There are three main differences to our preliminary evaluation. We utilise a more current version of the ASP solver Clingo, the Module Property is satisfied automatically, and we evaluated the influence of adding *External Statements* to our ASP program.

6.1   Dynamic Changes

The evaluation scenario has been modelled in two different ways. The first way, denoted as Ext, makes use of *External Statements* in its ASP rules, as described in Sect. 5. The second, denoted as noExt, purely relies on facts describing the connections between rooms of the department. Both ways utilize the transitive closure for their path planning approach, as presented in Listing 1.6. In Fig. 8 the robot's starting position is marked by a circle and the goal is marked by a cross. The path the robot is supposed to follow leaves the studentArea and follows mainHallA to reach mainHallB, since the door from mainHallA to the offices is closed. From mainHallB the path will enter the offices through a door and finally reaches the goal, which is situated in the utility area. This door is the solely open door in mainHallB and is opened and closed via an *External Statement* to simulate a change in the environment.

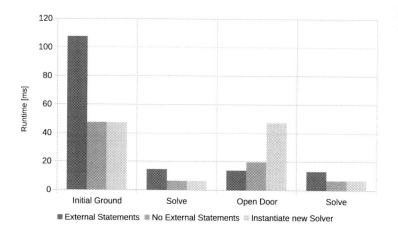

Fig. 9. Comparison of different modelling approaches [12].

The evaluation mainly consists of four steps. As a first step the solver is initialized and after transforming the ALICA plantree into corresponding ASP rules, the correctness of the plantree is checked as well as a navigation query is solved. In this step, the selected goal is not reachable, due to closed doors. The second step is purely solving the composed ASP program again, in order to be able to compare the time needed. In step three a door is opened, making the navigation between both rooms possible. This either means a new grounding step (noExt), the change of the truth value of an *External Statement* (Ext), or the creation of a new solver instance, which uses the noExt model. The last step is purely solving the ASP program again.

The evaluation results of our preliminary evaluation are presented in Fig. 9. In contrast to these former results, all steps were roughly halved in runtime, independent from the way changes are handled (see Fig. 10).

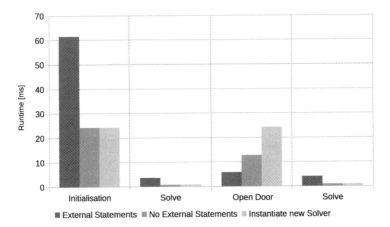

Fig. 10. Comparison of different modelling approaches using Automatic Module Property satisfaction.

The first and most time-consuming step lasts 61.5 ms for `Ext` and 24.2 ms for `noExt` or creating a new solver instance. The difference of 37.3 ms is caused by the way the department is modelled. In `noExt`, connections between rooms are expressed by single facts. In `Ext`, a connection between two rooms consists of an *External Statement* and two rules, which express the state of a door. This increases the size of the ASP program, and therefore, the initialization time. The second and the fourth step have similar results. Hereby, the usage of `Ext` results in an average runtime of 3.9 ms, while the usage of `noExt` results in an average runtime of 0.8 ms. In step three a door is opened or closed. This represents a dynamic change in the environment and is, therefore, more critical for our investigation than the other steps. The usage of `Ext` results in a runtime of 5.8 ms, `noExt` lasts 12.8 ms, and the instantiation of a new solver takes 24.2 ms. This is caused by the way a change in the model is performed. In `Ext`, the truth value of an *External Statement* is changed and thereby slightly influences the runtime. In `noExt`, a *Program Section* has to be grounded, which increases the runtime by roughly 12 ms. The highest increase of 23.6 ms is caused by a new solver instance since the initialization step has to be performed again.

As the ALICA framework usually runs at 30 Hz, an ALICA behaviour can query the ASP solver at most 30 times per second, i.e., each iteration. In Fig. 11, all three methods for modelling the department are compared with respect to the number of changes per 30 iterations.

The x-axis shows the number of changes per 30 iterations and the y-axis shows the average runtime of a query. The runtime for 30 changes per 30 queries, i.e., opening or closing a door each iteration, corresponds to the runtime of step three in Fig. 10. Since there is no difference in solving and changing a value when using `Ext` the blue line is constant. In comparison to this, the runtime of `noExt` increases when changes are made since a new program part has to be grounded for each change.

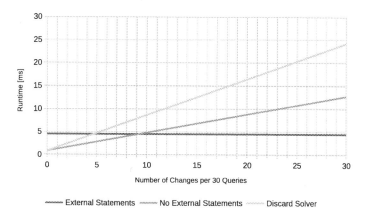

Fig. 11. Comparison of measured time regarding changes in the model.

As a result, *External Statements* are superior in runtime, when in 10 out of 30 iterations a door is closed or opened. This is due to the fact that, when using noExt, the stable models' sizes increase with every change made and slows down the process of handling the models rapidly. In contrast, when using Ext, the model's size stays the same. A third alternative is to discard the current and create a new ASPSolver instance after a few changes. This method is depicted by the green line, which intersects the blue line by six changes and is the slowest solution. Therefore, we suggest the use of Ext in dynamically changing environments.

6.2   Scalability of *External Statements*

Our approach to automatically satisfy the Module Property makes use of *External Statements*. Each time a query is formulated that was never queried before, a new *External Statement* is added to the ASP program. Although it is possible to deactivate old queries by releasing their corresponding *External Statements*, we encountered that old ASPExtensionQueries have a significant influence on the runtime. ASPFilterQueries do not suffer from this issue, as they do not alter the ASP program.

In Fig. 12 the runtime for an increasing number of *External Statements* is shown. It is important to note that there is always only one *External Statement* activated and all other *External Statements*, originating from former queries, are released. The average increment per additional *External Statement* is 0.1 ms. Therefore, the use of at most 1800 *External Statements* are allowed, without dropping below a query frequency of 5 Hz.

6.3   Scalability of the Region Connection Calculus

Following our preliminary evaluation, we re-evaluated the scalability of the Region Connection Calculus 4 using a more current version of Clingo. Utilising

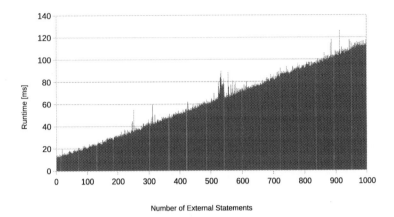

Fig. 12. Influence of increasing External Statements on the query runtime.

the navigation scenario from Sect. 4, we expanded the number of rooms, starting from 500 rooms and increasing up to 2000 rooms. Furthermore, we tested different connection densities between the rooms ranging from 25% to 100% connected rooms. Hereby, 100% means that every room has at least one connection to another room. The results are given in Figs. 13 and 14.

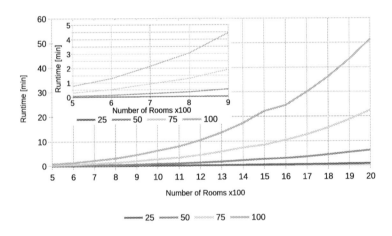

Fig. 13. Runtime of the initial grounding.

We stopped the evaluation at a number of 2000 rooms, as both runtimes increase exponentially with rising connection density and number of rooms. At 2000 rooms, the grounding already lasted 51 min. As a result, we only suggest using the Region Connection Calculus 4 in combination with transitive closure based path planning in dynamic domains if only less than 900 regions need to be considered. We consider this number suitable for dynamic domains since, after

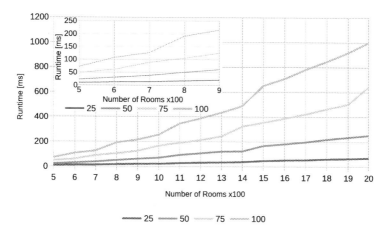

Fig. 14. Runtime of the solve step.

an initialization phase of roughly 4.5 min, a robot can still change the knowledge base 5 times per second by an ASPExtensionQuery.

7   Related Work

Besides the ALICA framework, other domain independent frameworks could be used to integrate an ASP solver for the use in different robotics scenarios. These frameworks include DyKnow [26] and KnowRob [27], which are both presented in this section. Furthermore, papers utilising ASP for multi-shot solving and the application of ASP in household scenarios are shown.

DyKnow, which is presented in [26], sets its focus on distributed collection and the distribution of data. This includes raw sensor values, processed sensor values or even predicates, which hold between objects recognised. Both, the collection and the distribution, are a set of processes specified in the knowledge processing language provided in [26]. These processes provide collected data, derived information, and knowledge about objects and their relation to multiple agents and allows them to reason about the received data. In comparison to this, the ALICA framework provides a domain independent framework, which is used to model and control the behaviour of agents. By expanding ALICA with the solver interface and the query mechanism presented, the agents are able to reason about the relations of objects in their environment and the domain specific knowledge, which is given in ALICA behaviours. Furthermore, ALICA provides the functionality to hierarchically constraint variables, which allows the formulation of queries utilising the ALICA plan tree structure.

KnowRob, presented in [27], is a framework utilising Prolog to build a knowledge base that provides access methods to retrieve information stored in the knowledge base. This knowledge base is extended by a "virtual knowledge base", which is used to compute the abstract representation of data when the data is queried. This is done by forwarding the query to another part of the robot, which has free resources or can provide better results. Furthermore, this framework supports only one agent, which is a contrast to ALICA which is used for teams of agents. In comparison to KnowRob framework, ALICA uses a declarative programming approach (ASP and Prolog), too. Additionally, both frameworks use this programming approach to formulate knowledge bases, which can be accessed by the supported agents.

Furthermore, in [8] one-shot solving is compared with multi-shot solving based on *External Statements*. Hereby, they used benchmarks given by the Fifth ASP Competition and support our results regarding *External Statements* in dynamic domains. Nevertheless, they always investigated *External Statements* in the context of expanding universes. According to our knowledge, our work is the first investigating the advantages of *External Statements* in the context of dynamic universes of almost constant size.

Erdem et al. present in [28] a framework utilising ASP and ConceptNet [29] for representing commonsense knowledge in ASP, which is then used to plan and execute household tasks. Hereby, ASP is used to represent the task of tidying a house consisting of three rooms that include a kitchen, a bathroom, and a living room. Therefore, the possible actions of a robot and the estimated locations of objects are modelled in ASP. This can be compared to our evaluation scenario (Sect. 5) and the application of the developed query mechanism. A household as presented in [28] is a highly dynamic and human populated environment and therefore is suited for the presented query mechanism, which has been proven is Sect. 6.

Finally, we want to state the difference to our preliminary work [12]. In [12] the user is required to manually guarantee the satisfaction of the Module Property. In this extended version, we introduce an approach to automatically guarantee the satisfaction of the Module Property. Whenever the user formulates an *ASPExtensionQuery* (in [12] denoted as *ASPVariableQuery*) our approach transforms the given query rules into unique rules without any interaction from a user. This makes the otherwise tedious satisfaction of the Module Property transparent to the user and releases him from this responsibility. Additionally, we re-evaluated our approach with a more current version (5.2.0) of the ASP solver Clingo that gives us a significant runtime improvement, as shown in Sect. 6.

8   Conclusion and Future Work

In extension to our preliminary work [12], we presented an automatic satisfaction of the Module Property and re-evaluated our scenarios with a current version of the ASP solver Clingo. The integration of Clingo with the ALICA framework clearly benefits from the current version of Clingo. All runtime evaluations show improved results. In order to prevent a violation of the Module Property, we automatically create a unique rule head for every part of the *ASPExtensionQuery*. As we showed before, both ways of creating unique rule heads (with or without *External Statements*) respond appropriately fast (less than 20 ms) for dynamic domains. Nevertheless, the runtime advantage with *External Statements* further improved, compared to the results in [12], due to the current Clingo version. We observed one disadvantage of *External Statements*. Whenever it is necessary to create a new query that is different from all other queries before, a new *External Statement* needs to be added to your ASP program. Although it is absolutely possible to do this automatically, we encountered an increase in solving time by 1 ms per created query. The amount of 1 ms depends on our modelling and can probably be further reduced, but the number of *External Statements*, even when they were already released, is a limiting factor with regard to scalability. However, we still propose the use of *External Statements* for modelling dynamic domains, such as human-populated service robotic domains. Only the number of different queries is limited to 1800 in order to remain reasonably low in runtime. As in [12], we evaluated the scalability of the Region Connection Calculus 4 by determining the transitive closure of the reachability relation. With the current Version of Clingo, we may conclude that the calculus scales up to a number of 900 instead of only 600 regions under premises that an agent can still query its knowledge base at a rate of 5 Hz after an initial grounding time of roughly five minutes.

In our future work, we will increase the variety of our scenarios in order to get a more profound impression of the validity of *External Statements* as a solution to reasoning in dynamic domains. Furthermore, this investigation will be joined with knowledge-based collaboration between multiple agents. In the current scenarios, the knowledge bases are independent of each other and the agents do not utilise knowledge from other agents knowledge bases. Implementing this feature, e.g., would allow an agent to ask another agent for open doors, instead of searching for open doors by itself.

Another aspect of knowledge-based collaboration is about global consistent stable models. Furthermore, it is possible in ASP that several valid models exist, but it is often desirable that a team agrees on the same or a similar model. Therefore, agents could exchange relevant parts or even complete models between each other, in order to choose the local model that is most similar to the parts received from other agents.

Besides the knowledge-based collaboration of agents, another part of our future work will be the provision of ASP based commonsense knowledge, which will enable agents to solve everyday tasks. A promising approach for this is the combination of Clingo with ConceptNet 5 [29], which represents commonsense knowledge as a hypergraph. This hypergraph consists of weighted edges connecting concepts with a given set of relations. These edges will be translated into ASP and will provide a commonsense knowledge base that can be accessed by an agent with the presented query structure.

References

1. Wurman, P.R., D'Andrea, R., Mountz, M.: Coordinating hundreds of cooperative, autonomous vehicles in warehouses. AI Mag. **29**, 9–19 (2008)
2. Hars, A.: Driverless Car Market Forecasts (2017). http://www.driverless-future. com/?page_id=384. 9 May 2017
3. Tomizawa, F.: Who is NAO? (2017). https://www.ald.softbankrobotics.com/en/ cool-robots/nao. 9 May 2017
4. Sullivan, A., Elkin, M., Umaschi Bers, M.: KIBO robot demo: engaging young children in programming and engineering. In: Proceedings of the 14th International Conference on Interaction Design and Children, pp. 418–421. ACM (2015)
5. van Overbeeke, B.: Service Robot Amigo at the RoboCup Dutch Open 2012 (2011). https://www.flickr.com/photos/robocup2013/9015099760. 19 May 2017
6. Brachman, R.J., Levesque, H.J.: Knowledge Representation and Reasoning. Morgan Kaufmann Series in Artificial Intelligence. Morgan Kaufmann, Los Altos (2003)
7. Gebser, M., Kaminski, R., Kaufmann, B., Schaub, T.: Clingo = ASP + Control: Extended Report (2014). https://www.cs.uni-potsdam.de/wv/publications/ TEMP_journals/corr/GebserKKS14x.pdf. 10 Oct 2017
8. Gebser, M., Janhunen, T., Jost, H., Kaminski, R., Schaub, T.: ASP solving for expanding universes. In: Calimeri, F., Ianni, G., Truszczynski, M. (eds.) Proceedings of the 13th International Conference on Logic Programming and Nonmonotonic Reasoning, pp. 354–367. Springer, Heidelberg (2015). https://doi.org/ 10.1007/978-3-319-23264-5_30
9. Skubch, H.: Modelling and Controlling of Behaviour for Autonomous Mobile Robots, 1st edn. Springer Vieweg, Heidelberg (2013). https://doi.org/10.1007/978-3-658-00811-6
10. Opfer, S., Niemczyk, S., Geihs, K.: Multi-agent plan verification with answer set programming. In: Aßmann, U., Brugali, D., Piechnick, C. (eds.) Proceedings of the Third Workshop on Model-Driven Robot Software Engineering. ACM (2016)
11. Gelfond, M., Kahl, Y.: Knowledge Representation, Reasoning, and the Design of Intelligent Agents: The Answer-Set Programming Approach. Cambridge University Press, Cambridge (2014)
12. Opfer, S., Jakob, S., Geihs, K.: Reasoning for autonomous agents in dynamic domains. In: van den Herik, J., Rocha, A.P., Filipe, J. (eds.) Proceedings of the 9th International Conference on Agents and Artificial Intelligence, pp. 340–351. ScitePress Digital Library (2017)
13. Randell, D.A., Cui, Z., Cohn, A.G.: A spatial logic based on regions and connection. In: Nebel, B., Rich, C., Swartout, W. (eds.) Proceedings of the 3rd International Conference on Principles of Knowledge Representation and Reasoning, vol. 92, San Francisco, CA, USA, pp. 165–176. Morgan Kaufmann (1992)
14. Skubch, H., Saur, D., Geihs, K.: Resolving conflicts in highly reactive teams. In: Luttenberger, N., Peters, H. (eds.) 17th GI/ITG Conference on Communication in Distributed Systems (KiVS 2011). OpenAccess Series in Informatics (OASIcs), vol. 17, pp. 170–175, Dagstuhl, Germany, Schloss Dagstuhl-Leibniz-Zentrum fuer Informatik (2011)
15. Skubch, H., Wagner, M., Reichle, R., Geihs, K.: A modelling language for cooperative plans in highly dynamic domains. In: van de Molengraft, M., Zweigle, O. (eds.) Special Issue on Advances in Intelligent Robot Design for the Robocup Middle Size League, vol. 21, pp. 423–433. Elsevier (2011)

16. Gat, E.: On three-layer architectures. In: Kortenkamp, D., Bonasso, R.P., Murphy, R. (eds.) Artificial Intelligence and Mobile Robots: Case Studies of Successful Robot Systems, pp. 195–210. MIT Press, Cambridge (1998)
17. Brewka, G., Eiter, T., Truszczyński, M.: Answer set programming at a glance. Commun. ACM **54**, 92–103 (2011)
18. Leone, N., Pfeifer, G., Faber, W., Eiter, T., Gottlob, G., Perri, S., Scarcello, F.: The DLV system for knowledge representation and reasoning. ACM Trans. Comput. Logic **7**, 499–562 (2006)
19. Eiter, T., Ianni, G., Krennwallner, T.: Answer set programming: a primer. In: Tessaris, S., Franconi, E., Eiter, T., Gutierrez, C., Handschuh, S., Rousset, M.-C., Schmidt, R.A. (eds.) Reasoning Web 2009. LNCS, vol. 5689, pp. 40–110. Springer, Heidelberg (2009). https://doi.org/10.1007/978-3-642-03754-2_2
20. Gebser, M., Grote, T., Kaminski, R., Obermeier, P., Sabuncu, O., Schaub, T.: Answer set programming for stream reasoning. In: Eiter, T., McIlraith, S. (eds.) Proceedings of the Thirteenth International Conference on Principles of Knowledge Representation and Reasoning (KR 2012), vol. 13, pp. 613–617. AAAI Press (2012)
21. Gatsoulis, Y., Alomari, M., Burbridge, C., Dondrup, C., Duckworth, P., Lightbody, P., Hanheide, M., Hawes, N., Hogg, D.C., Cohn, A.G.: QSRlib: a software library for online acquisition of qualitative spatial relations from video. In: Bredeweg, B., Kansou, K., Klenk, M. (eds.) Proceedings of the 29th International Workshop on Qualitative Reasoning, pp. 36–41 (2016)
22. Witsch, A.: Decision Making for Teams of Mobile Robots. Dissertation, University of Kassel, Kassel, Germany (2016)
23. Sharir, M.: A strong-connectivity algorithm and its applications in data flow analysis. Comput. Math. Appl. **7**, 67–72 (1981)
24. Schaub, T.: Module Composition, 13 October 2016. https://www.cs.uni-potsdam.de/~torsten/ijcai15tutorial/asp.pdf. 9 May 2017
25. Foote, T., Wise, M.: TurtleBot Website (2010). http://www.turtlebot.com/. 9 May 2017
26. Heintz, F., Kvarnström, J., Doherty, P.: Bridging the sense-reasoning gap: DyKnow - stream-based middleware for knowledge processing. Adv. Eng. Inform. **24**, 14–26 (2010)
27. Tenorth, M., Beetz, M.: KnowRob: a knowledge processing infrastructure for cognition-enabled robots. Int. J. Robot. Res. **32**, 566–590 (2013)
28. Erdem, E., Aker, E., Patoglu, V.: Answer set programming for collaborative housekeeping robotics: representation, reasoning, and execution. J. Intell. Serv. Robots **5**, 275–291 (2012)
29. Speer, R., Havasi, C.: ConceptNet 5: a large semantic network for relational knowledge. In: Gurevych, I., Kim, J. (eds.) The People's Web Meets NLP. Theory and Applications of Natural Language Processing, pp. 161–176. Springer, Heidelberg (2013). https://doi.org/10.1007/978-3-642-35085-6_6

Chest X-Ray Image Analysis

A Vision of Logic Programming

João Neves[1], Ricardo Faria[2], Victor Alves[2], Filipa Ferraz[2],
Henrique Vicente[2,3], and José Neves[2(✉)]

[1] Mediclinic Arabian Ranches, PO Box 282602, Dubai, United Arab Emirates
`joaocpneves@gmail.com`
[2] Centro Algoritmi, Universidade do Minho, Braga, Portugal
`ricardo.mof@hotmail.com,`
`{valves, jneves}@di.uminho.pt, filipatferraz@gmail.com`
[3] Departamento de Química, Escola de Ciências e Tecnologia,
Universidade de Évora, Évora, Portugal
`hvicente@uevora.pt`

Abstract. Most cardiovascular diseases can be prevented by addressing behavioral risk factors such as tobacco use, unhealthy diet and obesity, physical inactivity and harmful use of alcohol using strategies of the entire population. People with cardiovascular disease or high cardiovascular risk (due to the presence of one or more risk factors, such as hypertension, diabetes, hyperlipidemia or already established disease) need early detection and management using counseling and medication as appropriate. Now a leading cause of death. In fact, it reveals the centrality of prevention and how important it is to be aware of these situations. Thus, this paper will focus on the development of a decision support system to prevent these events to happen, centered on a structure based on Logic Programming for Representation and Knowledge Reasoning, complemented with a case-based approach to computation.

Keywords: Chest X-Ray images · Knowledge representation and reasoning
Logic programming · Case-based reasoning · Decision support systems

1 Introduction

The radiographic image is the result of using the X-Ray methods to visualize the inside of objects. These images are obtained with electromagnetic radiation in several wave ranges, which are selected accordingly to its usage. In the medical imaging field, the projectional radiography, as it is called, helps in diagnose, mainly the ones related to bones density or shape modifications. The chest X-Ray is a quick procedure, as well as easy, painless and non-invasive, and it allows to obtain images from the different structures within the chest area. With these images' records, it is possible to look for symptoms of several types of diseases, namely pneumonia, congestive heart failure, lung cancer, pulmonary fibrosis or sarcoid tissue.

© Springer International Publishing AG, part of Springer Nature 2018
J. van den Herik et al. (Eds.): ICAART 2017, LNAI 10839, pp. 48–61, 2018.
https://doi.org/10.1007/978-3-319-93581-2_3

In this study, the X-ray images will be used to evaluate cardiovascular problems, disease that cause 31.5% of the overall deaths in the world every year [1]. Indeed, this work is focused on the development of a hybrid methodology for problem solving, aiming at the elaboration of a decision support systems to detect cardiovascular problems based on parameters obtained from chest *X-ray* images, like the Cardiac Width (Fig. 1(a)), the Thoracic Width (Fig. 1(b)) and the *Aortic Knuckle Perimeter* (*AKP*) (Fig. 1(c)) [2], according to a historical dataset, under a *Case Based Reasoning* (*CBR*) approach to problem solving [3, 4].

Undeniably, *CBR* provides the ability of solving new problems by reusing knowledge acquired from past experiences [3], i.e., *CBR* is used especially when similar cases have similar terms and solutions, even when they have different backgrounds [4]. Its use may be found in many different arenas, like in *Online Dispute Resolution* [5] or *Medicine* [6–8], just to name a few.

This article is subdivided into five sections. In the former one a brief introduction to the problem is made. Then a mathematical logic approach to Knowledge Representation and Reasoning and a *CBR* view to computing are introduced. In the third and fourth sections a case study is set. Finally, in the last section the most relevant attainments are described and possible directions for future work are outlined.

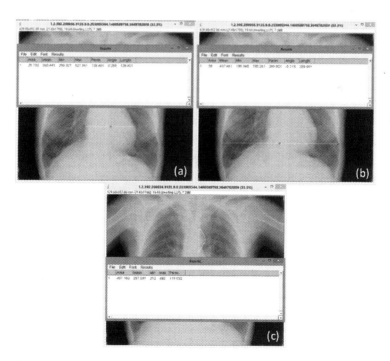

Fig. 1. The *Chest X-ray's* parameters considered in this study, i.e., *Cardiac Width* (a), *Thoracic Width* (b), and *Aortic Knuckle Perimeter* (c) [2].

2   Related Work

2.1   Knowledge Representation and Reasoning

Regarding the computational paradigm it were considered extended logic programs with two kinds of negation, classical negation, $\neg$, and default negation, *not* [9, 10].

An *Extended Logic Program* is a finite set of clauses as shown in Program 1.

$\{$

$$\neg p \leftarrow not\ p, not\ exception_p$$

$$p \leftarrow p_1 \wedge \cdots \wedge p_n \wedge not\ q_1 \wedge \cdots \wedge not\ q_m$$

$$?\left(p_1 \wedge \cdots \wedge p_n \wedge not\ q_1 \wedge \cdots \wedge not\ q_m\right)(n, m \geq 0)$$

$$exception_{p_1}$$

$$\cdots$$

$$exception_{p_j}\ (0 \leq j \leq k),\ being\ k\ an\ integer$$

$$\} :: scoring_{value}$$

Program 1. The Archetype of a Generic Extended Logic Program.

where the first clause of Program 1 depict the predicate's closure, "$\wedge$" denotes "*logical and*", while "*?*" is a domain atom denoting "*falsity*". The "p_i, q_j, and p" are classical ground literals, i.e., either positive atoms or atoms preceded by the classical negation sign "$\neg$" [9]. Indeed, "$\neg$" stands for a "*strong declaration*" that speaks for itself, and "*not*" denotes "*negation-by-failure*", i.e., a flop in proving a given statement, once it was not declared explicitly. According to this formalism, every program is associated with a set of "*abducibles*" [11, 12], given here in the form of *exceptions* to the extensions of the predicates that make the program, i.e., clauses of the form:

$$exception_{p_1}$$

$$\cdots$$

$$exception_{p_j}\ (0 \leq j \leq k),\ being\ k\ an\ integer$$

stands for data, information or knowledge that cannot be ruled out. The *invariants* or *restrictions*, i.e., clauses of the type:

$$?(p_1 \wedge \cdots \wedge p_n \wedge not\ q_1 \wedge \cdots \wedge not\ q_m)(n, m \geq 0)$$

allows one to set the context under which the universe of discourse has to be understood. Finally, the term *scoring*$_{value}$ denotes the relative weight of the extension of a specific predicate.

In order to model the universe of discourse in a changing environment, the breeding and executable computer programs will be ordered in terms of the *Quality-of-Information* (*QoI*) and *Degree-of-Confidence* (*DoC*) that stems out of them, when subject to a process of conceptual blending [13]. In blending, the structure or extension of two or more predicates is projected to a separate blended space, which inherits a partial structure from the inputs, and has an emergent structure of its own. Meaning is not compositional in the usual sense, and blending operates to produce understandings of composite functions or predicates, the conceptual domain, i.e., a basic structure of entities and relations at a high level of generality (e.g., the conceptual domain for journey has roles for traveler, path, origin, destination). Here it will be followed the normal view of conceptual metaphor, i.e., the system will carry structure from one conceptual domain (the source) to another (the target) directly. Now, the predicates whose extensions make an extended logic program (or theory that model the universe of discourse), i.e., i ($i \in \{1, ..., m\}$), the attributes of the mentioned predicates, j ($j \in \{1, ..., n\}$) and the respective values of the attributes j, i.e., $x_j \in [min_j, max_j]$ must be considered in order to compute a scoring function $V_j^i[min_j, max_j] \rightarrow 0 \cdots 1$, that gives the score predicate i assigns to a value of attribute j taking into account its domain, given in terms of *all* (*attribute exception list, sub expression, invariants*) productions. The former predicate generates a list of all possible value combinations (e.g., pairs, triples) as a list of sets defined by the domain size plus the invariants. The second predicate recourses through this list, and make a call to the third predicate for each exception combination. The third predicate denotes *sub expression* and is constructed in the same form. Thus, the *QoI* with respect to a generic predicate K is:

- 1 (one) when the information is *known* (*positive*) or *false* (*negative*);
- 0 (zero) if the information is *unknown*; and
- ϵ]0, 1[for situations where the extensions of the predicates that make the program include *exceptions* [14].

In order to measure the *QoI* that stems from a logic program or theory the *QoI* of each predicate are posting into a multi-dimensional space. The axes denote the logic program or theory, with a numbering ranging from 0 (at the center) to 1. Figure 2, shows an example of an extended logic program or theory P, built on the extension of 5 (five) predicates, $p_1 ... p_5$, where the dashed area stands for the respective *QoI*.

Regarding the *DoC*, it is a measure of one's confidence that the argument values of the terms that make the extension of a given predicate, with relation to their domains, fit into a given interval. The *DoC* is computed using $DoC = \sqrt{1 - \Delta l^2}$, where Δl denotes the length of the argument interval, which was set to the interval [0, 1], since the ranges

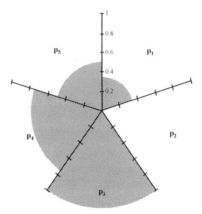

Fig. 2. A measure of *QoI* for the logic program or theory *P*.

of attributes values for a given predicate and respective domains were normalized, in terms of the expression $(Y - Y_{min})/(Y_{max} - Y_{min})$, where the Y_s stand for themselves.

The universe of discourse is engendered according to the information presented in the extensions of such predicates, according to productions of the type:

$$predicate_i - \bigcup_{1 \leq j \leq m} clause_j(((A_{x_1}, B_{x_1})(QoI_{x_1}, DoC_{x_1})), \cdots$$

$$\cdots, ((A_{x_n}, B_{x_n})(QoI_{x_n}, DoC_{x_n}))) :: QoI_j :: DoC_j$$

where $\cup$, *m* and *l* stand, respectively, for *set union*, the *cardinality* of the extension of *predicate_i* and the number of attributes of each clause [15]. On the other hand, either the subscripts of the QoI_s and the DoC_s, or those of the pairs (A_s, B_s), i.e., $x_1, ..., x_l$, stand for the attributes' clauses values ranges.

2.2 Case Based Computing

The CBR approach to computing is a technique for problem-solving grounded on the humans' decision-making process. Indeed, it stands for an act of finding and justifying a solution to a given problem based on the consideration of past similar situations, either using old solutions, or by reprocessing and generating new data or knowledge from the old ones [3, 4]. In *CB* the *cases* are stored in a *Case Base*, and those cases that are similar (or close) to a new one are used in the problem solving process. The typical *CB* cycle containing four steps, namely:

- Retrieve – The new case is defined and it is used to retrieve one or more cases from the *repository*, aiming to obtain cases with a higher degree of similarity to it;
- Reuse – The solutions of the similar cases found in the previous step were used to suggest a solution for the new problem;

- Revise – The suggested solution is tested by the user, allowing for its correction, adaptation and/or modification in order to found the solution for new problem; and
- Retain – The new problem and the correspondent solution is stored in the case *repository* [3, 4].

There are several examples of the use of CBR in Medicine, aiming to enhance the work of health experts and to improve the efficiency and quality of health care. In the literature it is possible to found review papers with different applications of *CBR* in medical context, such as disease diagnosis, classification, treatment and management [16, 17]. A recent study in the area of mental health care presents a *CBR* approach aiming to predict the effect of treatments of patients with anxiety disorders. Such approach showed 65% of correct predictions in the absence of similarity restrictions, while for scenario with similarity restrictions (i.e., under the condition that the prediction was based only on cases with a similarity of at least 0.62), the accuracy increased to 80% [6]. Another study presents a fuzzy ontology-based semantic *CBR* system to answer complex medical queries related to semantic understanding of medical concepts and handling of vague terms in diabetes diagnosis [7]. Other work combines *CBR* and multi-agent systems for the diagnosis, prognosis, treatment and therapeutic monitoring of gastric cancer. In the multi-agent architecture, the ontological agent type uses the knowledge domain in order to ensure proficiency in the extraction of similar clinical cases and provide treatment suggestions to patients and physicians. *CBR*, in turn, is used to memorize and to restore experience data aiming to solve similar problems [8].

Despite promising results, the current *CB* systems are neither complete nor adaptable enough for all domains. In some cases, the user cannot choose the similarity (ies) method(s) and is required to follow the system defined one(s), even if they do not meet their needs. Moreover, in real problems, the access to all necessary information is not always possible, since existent *CB* systems have limitations related to the capability of dealing, explicitly, with unknown, incomplete, and even self-contradictory information. To make a change, Neves *et al.* [2, 14] induced a different *CB* cycle that is depicted in Fig. 3. It takes into consideration the case's *QoI* and *DoC* metrics. It also contemplates a cases optimization process present in the *Case Base*, whenever they do not comply with the terms under which a given problem has to be addressed (e.g., the expected *DoC* on a prediction was not attained). The optimization process can use *Genetic Algorithms* [10], *Artificial Neural Networks* [18, 19] or *Particle Swarm Optimization* [20], generating a set of new cases which must be in conformity with the invariant:

$$\bigcap_{i=1}^{n} (\boldsymbol{B}_i, \boldsymbol{E}_i) \neq \emptyset \tag{1}$$

that states that the intersection of the attribute's values ranges for cases' set that make the *Case Base* or their optimized counterparts ($\boldsymbol{B}_i$) (being n its cardinality), and the ones that were object of a process of optimization ($\boldsymbol{E}_i$), cannot be empty (Fig. 3).

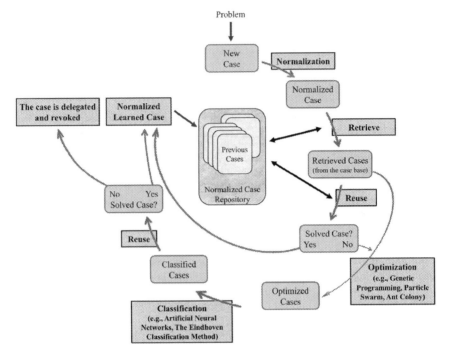

Fig. 3. The updated view of the *CB* cycle [2].

3 Case Study

Aiming to develop a predictive model to estimate the risk of cardiovascular diseases, a database was set, built on 542 health records of patients from a major health care institution in the North of Portugal. The patients included in this study aged between 18 to 97 years old, with an average of 56 ± 16 years old. The gender distribution was 42.2% and 57.8% for male and female, respectively.

After having collected the data it is possible to build up a knowledge database given in terms of the extensions of the relations or predicates depicted in Fig. 4, which stand for a situation where one has to manage information aiming to access the cardiovascular disease predisposing. The software *imageJ* [21] was used to extract the necessary features from *X-ray* images (Fig. 1). In the database some incomplete and/or default data is present. For instance, the *Systolic Blood Pressure* in case 1 are unknown (depicted by the symbol $\perp$), while the *Risk Factors* range in the interval [1, 2]. The *CTR* column is the *Cardiac Thoracic Ratio* computed using cardiac and thoracic width. The *Descriptions* column stands for free text fields that allow for the registration of relevant patient features.

Applying the algorithm presented in [2] to the fields that make the knowledge base for *Cardiovascular Diseases Predisposing* (Fig. 4), excluding at this stage of such a process the *Description* one, it is possible to set the arguments of the predicate *cardiovascular diseases predisposing* (*cdp*) referred to below, whose extension denote the objective function with respect to the problem under analyze:

Attributes of the Feature Vector:	#	Age	Systolic Blood Pressure	Cholesterol (LDL)	Cholesterol (HDL)	Triglycerides	Cardiac Thoracic Ratio	Aortic Knuckle Perimeter	Risk Factors	Description
					Cardiovascular Diseases Predisposing					
Feature Vector Attributes:	1	64	$\perp$	128	47	203	0.45	124	[1, 2]	Description 1
	2	72	128	135	53	252	$\perp$	$\perp$	[2, 3]	Description 2
	...	...	...	...	...	...	...	...	...	...
	542	44	113	92	45	104	0.42	112	0	Description 542
Feature Vector Domains:		[18, 97]	[70, 200]	[50, 250]	[20, 90]	[90, 600]	[0.35, 0.65]	[102, 155]	[0, 4]	

Fig. 4. A fragment of the knowledge base for cardiovascular diseases predisposing assessment.

$$cdp : Age, S_{ystolic}B_{lood}P_{ressure}, Chol_{esterol_{LDL}}, Chol_{esterol_{HDL}}, Trigly_{cerides},$$
$$C_{ardiac}T_{horacic}R_{atio}, A_{ortic}K_{nuckle}P_{erimeter}, R_{isk}F_{actors} \rightarrow \{0, 1\}$$

where 0 (zero) and 1 (one) denote, respectively, the truth values *false* and *true*.

Exemplifying the application of the algorithm presented in [2] to a term (patient) that presents the feature vector $Age = \perp$, $SBP = 120$, $Chol_{LDL} = 102$, $Chol_{HDL} = 70$, $Trigly = 130$, $CTR = [0.53, 0.56]$, $AKP = 120$, $RF = [1, 2]$, one may have:

$$\{$$

$$\neg \, cdp \left(\left((A_{Age}, B_{Age})(QoI_{Age}, DoC_{Age}) \right), \cdots, \left((A_{RF}, B_{RF})(QoI_{RF}, DoC_{RF}) \right) \right)$$

$$\leftarrow not \, cdp \left(\left((A_{Age}, B_{Age})(QoI_{Age}, DoC_{Age}) \right), \cdots, \left((A_{RF}, B_{RF})(QoI_{RF}, DoC_{RF}) \right) \right)$$

$$cdp \underbrace{\left(\left((0, 1)(1, 0) \right), \cdots, \left((0.25, 0.5)(1, 0.96) \right) \right)}_{\text{attribute's values ranges once normalized and respective QoI and DoC values}} :: 1 :: 0.86$$

$$\underbrace{[0, 1] \quad \cdots \quad [0, 1]}_{\substack{\text{attribute's domains} \\ \text{once normalized}}}$$

$$\} :: 1$$

4   Computational Model

The framework presented previously shows how the information comes together and how it is processed. In this section, a soft computing approach was set to model the universe of discourse, where the computational part is based on a *CB* approach to computing. In present work the *CB* cycle proposed by Neves *et al.* [2, 14] (Fig. 3) was adopted. The main advantage of the new *CB* cycle relies on the fact that not only all the cases have their arguments set in the interval [0, 1], but it also caters for the handling of incomplete, unknown, or even self-contradictory data or knowledge. Thus, the *Case Base* given in terms of the following pattern:

$$Case = \{ <Raw_{data}, \ Normalized_{data}, \ Description_{data} > \}$$

In addition the proposed methodology also contemplates the optimization of the retrieved cases. *Artificial Neural Networks* (*ANNs*) were used in the optimization stage in the following way:

- The extremes of the attribute's values ranges, as well as their *DoCs* and *QoIs* are fed to the *ANN*; and
- The outputs are given in a form that ensures that the case may be used to solve the problem (*no* (*0*), *yes* (*1*)), and a measure of the system confidence on such a result is provided (Fig. 5).

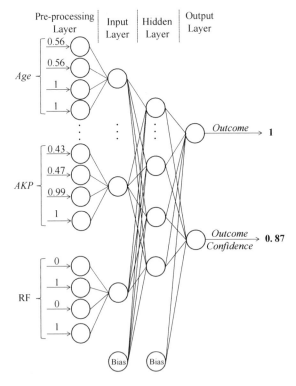

Fig. 5. A case's classification procedure based on *ANNs*.

When confronted with a new case, the system is able to retrieve all cases that meet such a case structure and optimize such a population, having in consideration that the cases retrieved from the *Case-base* must satisfy the invariant present in Eq. (1), in order to ensure that the intersection of the attributes range in the cases that make the *Case-base* repository or their optimized counterparts, and the equals in the new case cannot be empty. Having this in mind, the algorithm described in [2] is applied to the new case that presents the feature vector $Age = 62$, $SBP = 110$, $Chol_{LDL} = 105$, $Chol_{HDL} = 65$, $Trigly = 182$, $CTR = 0.43$, $AKP = [125, 127]$, $RF = \bot$, with the outcome:

$$cdp_{new\,case}(((0.56, 0.56)(1, 1)), \cdots, ((0, 1)(1, 0))) :: 1 :: 0.87$$

Now, the *new case* may be portrayed on the *Cartesian* plane in terms of its *QoI* and *DoC*, and by using clustering methods [22] it is feasible to identify the cluster(s) that intermingle with the *new one*. After the optimization process the *new case* is compared with every *retrieved case* from the cluster using a similarity function *sim*, given in terms of the average of the modulus of the arithmetic difference between the arguments of each case of the selected cluster and those of the *new case*. Thus, one may have:

$$retrieved_{case_1}((((0.77, 0.77)(1, 1)), \cdots, ((0.25, 0.75)(1, 0.87))) :: 1 :: 0.81$$
$$retrieved_{case_2}((((0.90, 0.90)(1, 1)), \cdots, ((0.5, 0.75)(1, 0.98))) :: 1 :: 0.85$$
$$\vdots$$
$$\underbrace{retrieved_{case_j}((((0.69, 0.69)(1, 1)), \cdots, ((0.75, 1)(1, 0.97))) :: 1 :: 0.83}_{normalized\ cases\ that\ make\ the\ retrieved\ cluster}$$

Assuming that every attribute has equal weight, for the sake of presentation, the *dis (imilarity)* between new_{case} and the $retrieved_{case1}$, i.e., $new_{case \to 1}$, may be computed as follows:

$$dis^{DoC}_{new\,case \to 1} = \frac{\|1 - 1\| + \cdots + \|1 - 0.87\|}{8} = 0.15$$

Thus, the *sim(ilarity)* for $sim^{DoC}_{new\,case \to 1}$ is set as $1 - 0.15 = 0.85$. Regarding *QoI* the procedure is similar, returning $sim^{QoI}_{new\,case \to 1} = 1$. Thus, one may have:

$$sim^{QoI,DoC}_{new\,case \to 1} = 1 \times 0.85 = 0.85$$

i.e., the product of two measurements is a new type of measurement. For instance, multiplying the lengths of the two sides of a rectangle gives its area, which is the subject of dimensional analysis. In this work the mentioned product gives the overall similarity between the new case and the retrieved ones. These procedures should be applied to the remaining cases of the retrieved clusters in order to obtain the most similar ones, which may stand for the possible solutions to the problem. This approach allows users to define the most appropriate similarity methods to address the problem (i.e., it gives the user the possibility to narrow the number of selected cases with the increase of the similarity threshold).

The proposed model was tested on a real data set with 542 examples. Thus, the dataset was divided in exclusive subsets through the ten-folds cross validation [19]. In the implementation of the respective dividing procedures, ten executions were performed for each one of them. Table 1 presents the coincidence matrix of the *CB*

Table 1. The coincidence matrix for *CB* model.

Target	Predictive	
	True (1)	False (0)
True (1)	276	33
False (0)	19	214

model, where the values presented denote the average of 30 (thirty) experiments. A perusal to Table 1 shows that the model accuracy was 90.4% (i.e., 490 instances correctly classified in 542). Thus, from clinical practice perspective, the predictions made by the *CB* model are satisfactory, attaining accuracies close to 90%.

Based on coincidence matrix it is possible to compute different metrics in order to evaluate the performance of the model, namely *sensitivity*, *specificity*, *Positive Predictive Value* (*PPV*), *Negative Predictive Value* (*NPV*) [23]. The sensitivity and specificity of the model were 89.3% and 91.8%, while *Positive* and *Negative Predictive Values* were 93.6% and 86.6%. The *ROC* curve [23, 24] is shown in Fig. 6. The area under *ROC* curve (0.91) denotes that the model exhibits a good performance in the assessment of cardiovascular diseases predisposing, despite the presence of unknown and incomplete information in the knowledge database.

In some recent studies the problem of incomplete information was addressed. In this context, Abreu et al. [25] present a study to predict the overall survival of women with breast cancer using a clinical dataset with 847 cases and 25% missing values. The *k*-nearest neighbor algorithm was used as the imputation method. The model presents a prediction accuracy of 73%. In another study the referred authors compared the performance of three different imputation methods, i.e., mean/mode imputation, expectation-maximization algorithm and *k*-nearest neighbor algorithm. The sensitivity, specificity and accuracy range from 83.9% to 88.4%, 47.4% to 70.5% and 68.8% to 81.7%, respectively [26].

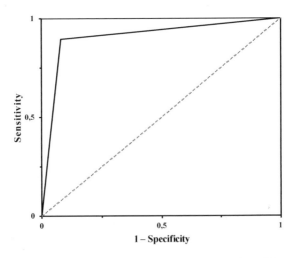

Fig. 6. The *ROC* curve regarding the proposed model.

5 Conclusion

CVD is one of the leading causes of death in the world. Adjusting lifestyle to minimize the prevalence of the risk factors can decrease the prevalence of CVD. Beyond just harming the individual who suffers from some form of *CVD*, it also affects everyone economically, environmentally, and socially. As healthcare and related costs increase, so do the budgets required by these programs. Thus, taxes will likely rise in order to keep up with the demand for government funding of these initiatives. Thus, this work aims at to minimize the impact of this situation, by presenting a *Logic Programming* based *Decision Support System* centered on a formal framework based on *LP* for knowledge representation and reasoning, complemented with a *CB* approach to computing that caters for the handling of incomplete, unknown, or even contradictory information. The proposed model is able to provide adequate responses once the overall accuracy is close to 90%. Indeed, it has also the potential to be disseminated across other prospective areas, therefore validating a universal attitude. In fact, the added values of the presented approach arises from the complementarily between *Logic Programming* (for knowledge representation and reasoning) and the computational process based on *Case Based* expertise.

Acknowledgments. This work has been supported by COMPETE: POCI-01-0145-FEDER-007043 and FCT – Fundação para a Ciência e Tecnologia within the Project Scope: UID/CEC/00319/2013.

References

1. Townsend, N., Wilson, L., Bhatnagar, P., Wickramasinghe, K., Rayner, M., Nichols, M.: Cardiovascular disease in Europe: epidemiological update 2016. Eur. Heart J. **37**, 3232–3245 (2016)
2. Faria, R., Alves, V., Ferraz, F., Neves, J., Vicente, H., Neves, J.: A case base approach to cardiovascular diseases using chest X-ray image analysis. In: van den Herik, J., Rocha, A.P., Filipe, J. (eds.) Proceedings of the 9th International Conference on Agents and Artificial Intelligence - ICAART 2017, vol. 2, pp. 266–274. Scitepress – Science and Technology Publications, Lisbon (2017)
3. Aamodt, A., Plaza, E.: Case-based reasoning: foundational issues, methodological variations, and system approaches. AI Commun. **7**, 39–59 (1994)
4. Richter, M.M., Weber, R.O.: Case-Based Reasoning: A Textbook. Springer, Heidelberg (2013). https://doi.org/10.1007/978-3-642-40167-1
5. Carneiro, D., Novais, P., Andrade, F., Zeleznikow, J., Neves, J.: Using case-based reasoning and principled negotiation to provide decision support for dispute resolution. Knowl. Inf. Syst. **36**, 789–826 (2013)
6. Janssen, R., Spronck, P., Arntz, A.: Case-based reasoning for predicting the success of therapy. Expert Syst. **32**, 165–177 (2015)
7. El-Sappagh, S., Elmogy, M., Riad, A.M.: A fuzzy-ontology oriented case-based reasoning framework for semantic diabetes diagnosis. Artif. Intell. Med. **65**, 179–208 (2015)

8. Shen, Y., Colloc, J., Jacquet-Andrieu, A., Lei, K.: Emerging medical informatics with case-based reasoning for aiding clinical decision in multi-agent system. J. Biomed. Inform. **56**, 307–317 (2015)

9. Neves, J.: A logic interpreter to handle time and negation in logic databases. In: Muller, R., Pottmyer, J. (eds.) Proceedings of the 1984 annual conference of the ACM on the 5th Generation Challenge, pp. 50–54. Association for Computing Machinery, New York (1984)

10. Neves, J., Machado, J., Analide, C., Abelha, A., Brito, L.: The halt condition in genetic programming. In: Neves, J., Santos, M.F., Machado, J.M. (eds.) EPIA 2007. LNCS (LNAI), vol. 4874, pp. 160–169. Springer, Heidelberg (2007). https://doi.org/10.1007/978-3-540-77002-2_14

11. Kakas, A., Kowalski, R., Toni, F.: The role of abduction in logic programming. In: Gabbay, D., Hogger, C., Robinson, I. (eds.) Handbook of Logic in Artificial Intelligence and Logic Programming, vol. 5, pp. 235–324. Oxford University Press, Oxford (1998)

12. Pereira, L.M., Anh, H.T.: Evolution prospection. In: Nakamatsu, K., Phillips-Wren, G., Jain, L.C., Howlett, R.J. (eds.) New Advances in Intelligent Decision Technologies. Studies in Computational Intelligence, vol. 199. Springer, Heidelberg (2009). https://doi.org/10.1007/978-3-642-00909-9_6

13. Turner, M., Fauconnier, G.: Conceptual integration and formal expression. J. Metaphor Symbolic Act. **10**, 183–204 (1995)

14. Esteves, M., et al.: Waiting time screening in diagnostic medical imaging – a case-based view. In: Tan, Y., Shi, Y. (eds.) Data Mining and Big Data. DMBD 2016. Lecture Notes in Computer Science, vol. 9714. Springer, Cham (2016). https://doi.org/10.1007/978-3-319-40973-3_30

15. Fernandes, F., Vicente, H., Abelha, A., Machado, J., Novais, P., Neves J.: Artificial neural networks in diabetes control. In: Proceedings of the 2015 Science and Information Conference (SAI 2015), pp. 362–370, IEEE Edition (2015)

16. Begum, S., Ahmed, M.U., Funk, P., Xiong, N., Folke, M.: Case-based reasoning systems in the health sciences: a survey of recent trends and developments. IEEE Trans. Syst. Man Cybern. Part C (Applications and Reviews) **41**, 421–434 (2011)

17. Blanco, X., Rodríguez, S., Corchado, Juan M., Zato, C.: Case-based reasoning applied to medical diagnosis and treatment. In: Omatu, S., Neves, J., Rodriguez, Juan M.Corchado, Paz Santana, J.F., Gonzalez, S.R. (eds.) Distributed Computing and Artificial Intelligence. AISC, vol. 217, pp. 137–146. Springer, Cham (2013). https://doi.org/10.1007/978-3-319-00551-5_17

18. Vicente, H., Dias, S., Fernandes, A., Abelha, A., Machado, J., Neves, J.: Prediction of the quality of public water supply using artificial neural networks. J. Water Supply Res. Technol. AQUA **61**, 446–459 (2012)

19. Haykin, S.: Neural Networks and Learning Machines. Pearson Education, New Jersey (2009)

20. Mendes, R., Kennedy, J., Neves J.: Watch thy neighbor or how the swarm can learn from its environment. In: Proceedings of the 2003 IEEE Swarm Intelligence Symposium (SIS 2003), pp. 88–94, IEEE Edition (2003)

21. Rasband, W.S.: ImageJ. U. S. National Institutes of Health, Bethesda, Maryland, USA 1997–2015. http://imagej.nih.gov/ij/

22. Figueiredo, M., Esteves, L., Neves, J., Vicente, H.: A data mining approach to study the impact of the methodology followed in chemistry lab classes on the weight attributed by the students to the lab work on learning and motivation. Chem. Educ. Res. Pract. **17**, 156–171 (2016)

23. Florkowski, C.M.: Sensitivity, specificity, Receiver-Operating Characteristic (ROC) curves and likelihood ratios: communicating the performance of diagnostic tests. Clin. Biochem. Rev. **29**(Suppl 1), S83–S87 (2008)
24. Hajian-Tilaki, K.: Receiver Operating Characteristic (ROC) curve analysis for medical diagnostic test evaluation. Caspian J. Intern. Med. **4**, 627–635 (2013)
25. Abreu, P., Amaro, H., Silva, D., Machado, P., Abreu, M., Afonso, N., Dourado, A.: Overall survival prediction for women breast cancer using ensemble methods and incomplete clinical data. In: Romero, L. (ed.) XIII Mediterranean Conference on Medical and Biological Engineering and Computing 2013, IFMBE Proceedings, vol. 41, pp. 1366–1369. Springer, Switzerland (2014). https://doi.org/10.1007/978-3-319-00846-2_338
26. García-Laencina, P., Abreu, P., Abreu, M., Afonso, N.: Missing data imputation on the 5-year survival prediction of breast cancer patients with unknown discrete values. Comput. Biol. Med. **59**, 125–133 (2015)

Text Classification and Transfer Learning Based on Character-Level Deep Convolutional Neural Networks

Minato Sato[✉], Ryohei Orihara[✉], Yuichi Sei[✉], Yasuyuki Tahara[✉],
and Akihiko Ohsuga[✉]

Graduate School of Information Systems, The University of Electro-Communications,
1-5-1, Chofu-gaoka, Chofu-shi, Tokyo, Japan
sato.minato@ohsuga.is.uec.ac.jp, ryohei.orihara@toshiba.co.jp,
{seiuny,tahara,ohsuga}@uec.ac.jp

Abstract. Temporal (one-dimensional) Convolutional Neural Network (Temporal CNN, ConvNet) is an emergent technology for text understanding. The input for the ConvNets could be either a sequence of words or a sequence of characters. In the latter case there are no needs for natural language processing. Past studies showed that the character-level ConvNets worked well for text classification in English and romanized Chinese corpus. In this article we apply the character-level ConvNets to Japanese corpus. We confirmed that meaningful representations are extracted by the ConvNets in English corpus and Japanese corpus. We attempt to reuse the meaningful representations that are learned in the ConvNets from a large-scale dataset in the form of transfer learning. As for the application to the news categorization and the sentiment analysis tasks in Japanese corpus, the ConvNets outperformed N-gram-based classifiers. In addition, our ConvNets transfer learning frameworks worked well for a task which is similar to one used for pre-training.

Keywords: Deep learning · Temporal ConvNets · Transfer learning
Text classification · Sentiment analysis

1 Introduction

Recently, many deep learning algorithms have achieved high accuracy in media information processing, such as image and speech recognition. Concretely, a deep convolutional neural network (ConvNet, CNN) achieved a winning top-5 test error rate of 15.3%, compared to 26.2% achieved by the second-best entry in ILSVRC-2012 competition [10]. Many deep learning approaches in the field of image recognition reuse the ConvNets pre-trained on a very large dataset (e.g. ImageNet [4]) as a parameter initialization or feature extractor [23]. This approaches are called transfer learning, in particular, inductive transfer.

A successful example of deep learning in the field of natural language processing is word2vec [15,16] which is a method for producing word embeddings (*i.e.*, low-dimensional and dense vectors).

© Springer International Publishing AG, part of Springer Nature 2018
J. van den Herik et al. (Eds.): ICAART 2017, LNAI 10839, pp. 62–81, 2018.
https://doi.org/10.1007/978-3-319-93581-2_4

In the field of natural language processing, text classification is a classic and important task because of many applications. It was important for classic text classification to make "good" features for a classifier by hand. For instance, a language-specific sentiment dictionary for sentiment analysis is often employed and it needs a language-specific morphological analysis to make a Bag-of-Words model or a Bag-of-N-grams model.

There are deep learning approaches for text classification/sentiment analysis. In the early days, stacked denoising autoencoders (SDAs) are used for Amazon review data sentiment analysis [7]. Recursive neural networks (RNNs) based on a syntax tree are also used for a similar task [24]. A syntactic analyzer or a manpower is needed to generate the syntax tree.

Recent approaches are to apply the temporal ConvNets on a sequence of words or a sequence of characters. The former, the word-level ConvNets require the morphological analyzer of a target language. The word embeddings pre-trained by unsupervised learning (*i.e.*, word2vec training) improve classification accuracy of the word-level ConvNets. The latter, the character-level ConvNets do not require the morphological analyzer, the syntactic analyzer, etc. Past studies showed that the character-level ConvNets worked well for news category classification and sentiment analysis/classification tasks in English and romanized Chinese text corpus. However, any of other languages has not been studied yet. Additionally, nobody discusses features extracted by the character-level ConvNets although many discussions about that are in the field of image recognition.

This study experiments on Japanese text classification with the ConvNets, and we investigate what the character-level ConvNets extract and the possibility of transfer learning in order to make good use of them.

The remainder of this paper is organized as follows. Section 2 explains related works dealing with the temporal ConvNets for text classification and an overview of transfer learning in the field of image recognition. Section 3 briefly describes the character-level ConvNets and the transfer learning frameworks for experiments. Section 4 provides the dataset description and the experimental results. Section 5 provides the discussions for the experimental results. Finally, Sect. 6 gives a conclusion of this paper.

2   Related Work

2.1   Word-Level ConvNet

The word-level ConvNets approaches which apply the temporal ConvNets on the sequence of words are proposed by Kim [9] and Severyn et al. [21, 22]. Kim uses multiple window sizes of convolution filters and experiments on movie sentiment analysis. Severyn et al. experiment on very short text, such as Twitter and SMS. Both of their models use only one temporal convolutional layer and one temporal pooling layer, namely they are relatively shallow.

2.2   Character-Level ConvNet

The character-level ConvNets approaches which apply the temporal ConvNets on the sequence of characters are proposed by Santos et al. [18,19] and Zhang et al. [26,27]. Santos et al. combine the word-level ConvNet with a pre-trained embedding layer and the character-level ConvNet with a randomly initialized embedding layer for a twitter sentiment analysis. Their model is also relatively shallow, like past studies of the word-level ConvNets. Zhang et al. are the first to apply the temporal ConvNets only on the sequence of characters. An acceptable input of their model is the sequence of one-hot (or 1-of-m) encoded characters. In the case of English dataset, the dimensionality of one-hot encoding is the sum of the number of letters from "a" to "z", digits "0" to "9" and signs (*e.g.*, "?", "!", etc.), hence at most 70. In the case of Chinese dataset, the past study uses the romanization by Python library `pypinyin`. Hence, the dimensionality of one-hot encoding for Chinese input is the same as in the case of English dataset. Their model is a deep architecture which is composed of 6 convolutional-pooling layers and 3 fully-connected layers.

2.3   Transfer Learning Based on ConvNet in Image Recognition

In image recognition of deep learning, many researchers study reuse of the ConvNets pre-trained on ImageNet [4]. Razavian et al. reuse the pre-trained ConvNets as a feature extractor [23], and they apply linear SVM classifier on extracted features. One of other approaches is replacing weights of the output layer of the pre-trained ConvNets with randomly initialized weights and re-training the whole weights of the ConvNets on a new task [1,5]. This approach is called fine-tuning. In general, accuracy of the ConvNets has a strong tendency to depend on the initial weights. Weight initialization by the pre-trained ConvNets produces better results than random weight initialization. In particular, if the number of samples of a training dataset is small, good initial weights prevent the ConvNets from overfitting and enhance a generalization ability of the ConvNets.

3   Character-Level ConvNet

3.1   Overview

This section shows key modules for applying the temporal ConvNets on the character-level input [18,19,26,27].

Input Representation. We employ two types of input representation for experiments. One is a simple one-hot representation which follows the past study. Another is a distributed representation (*i.e.*, a character-level embedding) in order to omit troublesome romanization processing in the case of Japanese dataset.

Given a sentence composed of N characters $\{c_1, c_2, \cdots, c_N\}$, we first transform each character c_n into the d-dimensional one-hot representation which has value 1 at index c_n and zero in all other positions. Hence, the sentence composed of the N characters is transformed into $\{r_1, r_2, \cdots, r_N\} \in \mathbb{R}^{d \times N}$.

Character-level embeddings require an additional transformation. The character-level embeddings are encoded by column vectors in an embedding matrix $W^e \in \mathbb{R}^{d^e \times d}$ which is a parameter to be learned. We transform each one-hot character vector r_n into the d^e-dimensional character-level embedding $r_n^e \in \mathbb{R}^{d^e}$ by using the matrix-vector product:

$$r_n^e = W^e r_n. \tag{1}$$

The dimensionality of the character-level embedding d^e is a hyper-parameter to be chosen by the user. In the case of the character-level embeddings, the sentence composed of the N characters is finally transformed into $\{r_1^e, r_2^e, \cdots, r_N^e\} \in \mathbb{R}^{d^e \times N}$.

In the following explanation, the vector s_n represents either the vector r_n or the vector r_n^e.

Temporal Convolution. A vector $z_n \in \mathbb{R}^{d \times k}$ applied to temporal convolution with a filter of window size k is defined as a concatenation of the one-hot representations:

$$z_n = \left(s_{n-(k-1)/2}, \cdots, s_{n+(k-1)/2}\right)^T. \tag{2}$$

Hence, the output value of temporal convolution by the ith filter is as follows:

$$[u_n]_i = [W z_n + b]_i \tag{3}$$

where $W \in \mathbb{R}^{f \times d \times k}$ is the weight matrix composed of the f convolution filters and $b \in \mathbb{R}^f$ is the bias vector. W and b are parameters to be learned.

Temporal Pooling. Let us define $M = N - (k-1)$ and temporal max-pooling size is p. The output value of temporal convolution by the ith filter can be describe as $(v_1, v_2, \cdots, v_M)_i \in \mathbb{R}^{N-(k-1)}$. Hence, an applied range of temporal max-pooling $[y_m]_i$ is as follows:

$$[y_m]_i = \left(v_{m-(p-1)/2}, \cdots, v_{m+(p-1)/2}\right)_i. \tag{4}$$

The dimensionality of the output matrix of the temporal max-pooling layer is $f \times (N - (k-1))/p$.

3.2   Model Design

Below we prepare shallow ConvNets and deep ConvNets. The sequential information in the text whose length is larger than the window size, cannot be taken into consideration with the shallow ConvNets. In the meantime, multiple convolution and pooling enable the deep ConvNets to take those information into consideration.

The alphabet set used in the models of one-hot representation consists of the following 68 characters.

```
abcdefghijklmnopqrstuvwxyz0123456789
,;.!?:'/\|_@#$%^&*~'"+-=<>(){}
```

The input sentence length (*i.e.*, the number of input characters) is fixed to 1014. Hence, in the case of the one-hot representation, the dimensionality of the input matrix is 68×1014. The dimensionality of the character-level embedding is set to 50. Hence, in the case of the character-level embedding, the dimensionality of the input matrix is 50×1014. In addition, ReLU [17] is applied to all the layers except for the output layer. For the network training, momentum SGD [2] is carried out via backpropagation. Then, the mini-batch size is set to 50 and the momentum is set to 0.9 and initial learning rate is set to 0.01 which is halved every 3 epochs for 10 times. The weights of all the models are initialized by "Xavier initialization" [6].

Deep Model. Zhang et al. designed two models of 9 layers deep neural networks with 6 convolutional layers and 3 fully-connected layers. One of the two models has a large number of the convolution filters (Large-C6FC3), while another one has a small number of those (Small-C6FC3). Cn_1FCn_2 refers to a network with n_1 convolutional layers and n_2 fully-connected layers. Tables 1 and 2 list the configurations of the deep models of Zhang et al. The former lists the configurations of the 6 convolutional layers of the deep models and the latter lists the configurations of the 3 fully-connected layers of the deep models. Each column of Table 1 indicates the index of the layers, the number of convolution filters of the large model, the number of convolution filters of the small model, the window size of the convolution filters and the max-pooling size. Each column of Table 2 indicates the index of the layers, the number of the output units of the large model and the number of the output units of the small model. The 9th layer is

Table 1. Configurations of the 6 convolutional layers of Small-C6FC3 and Large-C6FC3 [20].

Layer	Large frame	Small frame	Window	Pool
1	1024	256	7	3
2	1024	256	7	3
3	1024	256	3	N/A
4	1024	256	3	N/A
5	1024	256	3	N/A
6	1024	256	3	3

Table 2. Configurations of the 3 fully-connected layers of Small-C6FC3 and Large-C6FC3 [20].

Layer	Output units large	Output units small
7	2048	1024
8	2048	1024
9	Depends on the problem	

the output layer, hence the number of units of that layer depends on a problem at hand (*e.g.*, if we deal with a sentiment polarity classification problem, the number of those is two.). To regularize the network, the dropout method [25] is applied to be between the 3 fully-connected layers with a probability of 0.5.

Shallow Model. We build the shallow character-level ConvNets for comparing with the deep models of Zhang et al. Tables 3 and 4 list the configurations of the shallow models like Tables 1 and 2.

Table 3. Configurations of the convolutional layer of Small-C1FC1 and Large-C1FC1 [20].

Layer	Large frame	Small frame	Kernel	Pool
1	1024	256	7	1008

Table 4. Configurations of the fully-connected layer of Small-C1FC1 and Large-C1FC1 [20].

Layer	Output units large	Output units small
2	Depends on the problem	

3.3 Character-Level ConvNets for Transfer Learning

A target task in this section is a text classification task in a relatively small dataset. To prevent the ConvNets from overfitting, we train the ConvNets on a very large dataset and reuse them with transfer learning frameworks.

We employ Small-C6FC3 model for experiments dealing with transfer learning. The following three transfer learning frameworks are compared each other:

Scratch. The model without pre-training on the very large dataset.

Pre-trained Feature. We initialize the 6 convolutional layers of the model with those of the pre-trained model, and freeze those layers for training. Thus, the pre-trained model is used for the feature extractor and only the 3 fully-connected layers are trained through backpropagation.

Fine-tuning. We initialize all the layers without the output layer of the model with those of the pre-trained model and do not freeze any of the layers for training.

4   Experiments

4.1   Baseline Methods

We use a Bag-of-Words model and a Bag-of-N-grams model as baseline methods. We employ a multinomial logistic regression for tf-idf features based on a dictionary created by the following methods.

Bag-of-Words. In the case of English dataset, pre-processing (*e.g.*, removing stopwords) is carried out by Python library `gensim`. In the case of Japanese dataset, Japanese morphological analysis is carried out by `MeCab` [11]. For both of English dataset and Japanese dataset, we use the most frequent 5000 words as the dictionary.

Bag-of-N-grams. In the case of a English dataset, we use the most frequent n-grams (up to 5-g) as the dictionary. In the case of Japanese dataset, Japanese romanization is carried out by "Kanji Kana Simple Inverter" (`KAKASI`[1]). We use the most frequent n-grams (up to 5-g) that are romanized as the dictionary.

4.2   Datasets and Results for Japanese Text Classification

This study employs two types of task for each of English and Japanese with reproduction of the past study in mind. One of the tasks is news categorization, while another one is sentiment analysis.

AFPBB Dataset. We collected Japanese news articles including titles, texts and categories from AFPBB News[2] for the Japanese news categorization task. Table 5 describes an overview of the corpus. The corpus contains 79,778 articles from May 2006 to May 2016. The categories of the corpus are composed of "Lifestyle", "Politics", "Science" and "Sports". We sampled 12,000 articles for each of the categories as a training dataset while we sampled 500 articles for each of the categories as a validation dataset and a test dataset. The romanization is carried out for the input for the Bag-of-N-grams and the one-hot character-level ConvNet.

The results for the AFPBB dataset are shown in Table 6. From the point of view of classification accuracy, the best model is the Bag-of-Words model. The results indicate that C1FC1 one-hot models outperform C6FC3 one-hot models, and C6FC3 embedding models outperform C1FC1 embedding models. We suppose C6FC3 one-hot models fall into overfitting because their expressive power derived from the deep architecture is too much for the small AFPBB dataset and the deteriorated input information due to the romanization. As for C6FC3 embedding models, we suppose that the intact input information and the relatively deep architecture improve the accuracy. However, none of the ConvNets outperform the Bag-of-Words model because they fail to learn features as good as the dictionary from the relatively small dataset.

---

[1] http://kakasi.namazu.org/.

[2] http://www.afpbb.com/.

Table 5. Overview of AFPBB dataset [20].

Category	Total	Train	Validation	Test
Lifestyle	21,927	12,000	500	500
Politics	18,221	12,000	500	500
Science	13,069	12,000	500	500
Sports	26,561	12,000	500	500

Table 6. Classification accuracy for AFPBB dataset [20].

Model	Accuracy
Bag-of-Words	**0.947**
Bag-of-N-grams	0.926
Small-C1FC1 one-hot	0.9385
Large-C1FC1 one-hot	0.941
Small-C1FC1 embedding	0.916
Large-C1FC1 embedding	0.9225
Small-C6FC3 one-hot	0.9120
Large-C6FC3 one-hot	0.9295
Small-C6FC3 embedding	0.9365
Large-C6FC3 embedding	0.9395

Rakuten Market Review Dataset. We obtained a Rakuten[3] market review dataset from the Informatics Research Data Repository of National Institute of Informatics[4]. We created a subset of it by extracting only reviews with ★1 and ★5 and written between April 2012 and December 2012 to use in the Japanese sentiment analysis task. Among sentiment analysis tasks, this study deals with a sentiment polarity classification (a task for classifying a input text into positive or negative). We sampled 680,000 reviews from the Rakuten market review dataset as shown in Table 7. We created the dictionary for the Bag-of-Words and the Bag-of-N-grams from 200,000 reviews which are randomly sampled from the training dataset because of the limitations of memory.

Table 8 shows the results. We had two assumptions. One is that the sequential information improves the accuracy for the sentiment analysis task. Another assumption is that the words or the N-grams are more important than the sequential information in the news categorization task. However, simple C1FC1 models are not inferior to C6FC3 models which is able to take the sequential information into consideration. On the contrary, one of the shallow models, Large-C1FC1 with the character-level embedding is the best model. C6FC3

---

[3] Rakuten, Inc. is one of the largest Japanese electronic commerce and Internet companies based in Tokyo, Japan.

[4] http://www.nii.ac.jp/dsc/idr/en/rakuten/rakuten.html.

Table 7. Overview of Rakuten market review dataset [20].

Polarity	Total	Train	Validation	Test
Positive(★5)	11,434,454	300,000	20,000	20,000
Negative(★1)	370,160	300,000	20,000	20,000

models outperform C1FC1 models in the case of the one-hot representation, while C1FC1 models outperform C6FC3 models in the case of the character-level embedding. Since a normal Japanese sentence is shorter than a romanized Japanese sentence, the applied range of convolution or max-pooling differs between the normal Japanese sentence and the romanized Japanese sentence. Thus, the Japanese character-level embedding enables the ConvNets to have more broader viewpoints than the romanized one-hot representation. We suppose that the broader viewpoints have provided important information in the sentiment analysis task, which usually comes from the sequential information.

Table 8. Classification accuracy for Rakuten market review dataset [20].

Model	Accuracy
Bag-of-Words	0.95475
Bag-of-N-grams	0.950975
Small-C1FC1 one-hot	0.958525
Large-C1FC1 one-hot	0.963725
Small-C1FC1 embedding	0.967675
Large-C1FC1 embedding	**0.969875**
Small-C6FC3 one-hot	0.9637
Large-C6FC3 one-hot	0.966825
Small-C6FC3 embedding	0.966925
Large-C6FC3 embedding	0.9689

AG News Dataset. We obtained AG's corpus of news articles [3,8] from Gulli's website[5] for English categorization task. We sampled 400,000 articles which belong to any of four largest categories from the corpus for the training dataset, the validation dataset and the test dataset as shown in Table 9. We created the dictionary for the Bag-of-Words and the Bag-of-N-grams from the random sampled 120,000 reviews because of the limitations of memory.

Table 10 shows the results. The best model is Large-C1FC1. There is, however, not much different between all the models, because all the models could extract key words as features.

---

[5] https://www.di.unipi.it/~gulli/AG_corpus_of_news_articles.html.

Table 9. Overview of AG news dataset [20].

Category	Total	Train	Validation	Test
World	186,674	90,000	5,000	5,000
Sports	118,103	90,000	5,000	5,000
Business	134,223	90,000	5,000	5,000
Sci/Tech	153,595	90,000	5,000	5,000

Table 10. Classification accuracy for AG news dataset [20].

Model	Accuracy
Bag-of-Words	0.8689
Bag-of-N-grams	0.87665
Small-C1FC1	0.8701
Large-C1FC1	**0.8849**
Small-C6FC3	0.87095
Large-C6FC3	0.87925

Amazon Review Dataset. We obtained an Amazon review dataset [13,14] including eight categories ("Books", "Electronics", etc.) from the Stanford Network Analysis Project (SNAP)[6] for the English sentiment analysis task. We sampled 760,000 reviews whose rating is ★1 or ★5 from this dataset for the training dataset, the validation dataset and the test dataset as shown in Table 11. We created the dictionary for the Bag-of-Words and the Bag-of-N-grams from the random sampled 200,000 reviews because of the limitations of memory.

Table 11. Overview of Amazon review dataset [20].

Polarity	Total	Train	Validation	Test
Positive(★5)	8,829,533	300,000	40,000	40,000
Negative(★1)	653,333	300,000	40,000	40,000

Table 12 shows the results. The best model is Large-C6FC3. Since the dictionary based on the frequent words may not contain important key words for the sentiment analysis, the flexible ConvNets are proven to be more accurate than the Bag-of-Words and the Bag-of-N-grams by a 3–5% margin.

---

[6] https://snap.stanford.edu/data/web-Amazon.html.

Table 12. Classification accuracy for Amazon review dataset [20].

Model	Accuracy
Bag-of-Words	0.882175
Bag-of-N-grams	0.881275
Small-C1FC1	0.91125
Large-C1FC1	0.925425
Small-C6FC3	0.92155
Large-C6FC3	**0.931925**

4.3   Datasets and Results for Transfer Learning

We construct English dataset and Japanese dataset for the experiment for transfer learning.

For the experiment for the English dataset, we construct a large-scale dataset including sixteen categories which are shown in Table 13 for pre-training from the Amazon review dataset. We call it a pre-training dataset. Table 14 shows an overview of the pre-training dataset. The test classification accuracy of Small-C6FC3 for the pre-training dataset is 0.9168.

Table 13. Sixteen categories for pre-training [20].

Category
Apps for Android
Automotive
Baby
Beauty
Books
Clothing shoes and jewelry
Digital music
Grocery and gourmet food
Health and personal care
Office products
Patio lawn and garden
Pet supplies
Sports and outdoors
Tools and home improvement
Toys and games
Video games

Table 14. Overview of Amazon review dataset for pre-training [20].

Overall	Polarity	Train	Validation	Test
★5	Positive	400,000	40,000	40,000
★4		200,000	20,000	20,000
★2	Negative	200,000	20,000	20,000
★1		400,000	40,000	40,000

We choose "Movies and TV", "Electronics", "Home and Kitchen" category to construct the small-scale datasets for the target task. We call it a target dataset. It should be noted that the pre-training dataset does not include these three categories. Table 15 shows an overview of the target dataset. The ratio of the number of the reviews for each of ★1, ★2, ★4 and ★5 rating is 2:1:1:2, namely this ratio is the same ratio as the large-scale dataset for the pre-training as shown in Table 14.

Table 15. Overview of the English dataset for main task of transfer learning. To be specific, "Movies" indicates "Movies and TV" category and "Home" indicates "Home and Kitchen" category [20].

	Movies	Electronics	Home
Train	150,000	150,000	60,000
Validation	30,000	30,000	9,000
Test	30,000	30,000	9,000

Table 16 shows the results. The results indicate that the Fine-tuning is the best method for all the datasets. Additionally, the smaller the dataset gets, the less accurate the Scratch framework gets. From these facts, making good use of the pre-trained ConvNets is very effective for the prevention of overfitting and the enhancement of the generalization ability.

Table 16. The results of experiments for transfer learning on the Amazon review dataset. Each of numbers indicates the classification accuracy [20].

Model\Dataset	Movies	Electronics	Home
Bag-of-Words	0.85503	0.86033	0.8524
Bag-of-N-grams	0.85603	0.8755	0.8693
Scratch	0.85827	0.87854	0.8581
Pre-trained feature	0.88697	0.88183	0.8988
Fine-tuning	**0.8992**	**0.90523**	**0.9123**

In addition, we used a IMDb review dataset [12] for the target dataset. The task of this dataset is also the sentiment polarity classification task. This dataset was collected 50,000 polarized reviews from the Internet Movie Database[7]. A negative review has a score 4 or less out of 10, while a positive review has a score 7 or more out of 10. The default training dataset has 25,000 reviews, while the default test dataset has 25,000 reviews. We randomly sampled 2,500 reviews for the validation dataset from the default training dataset. Hence, we could use only 22,500 reviews for the training dataset. A polarity ratio of all the datasets is 1:1. Since we have an assumption that a characteristic (*e.g.*, scale, a topic) of the pre-training dataset influences the results of transfer learning, we construct another pre-training dataset including only "Movies and TV" category which is closely related topic to the IMDb review dataset.

Table 17 shows the results. The Scratch framework falls into overfitting because of its depth and the smallness of the IMDb review dataset. The transfer learning frameworks using the pre-trained ConvNets outperform the classic text classification methods. Furthermore, the framework using the ConvNets pre-trained on the large-scale dataset including the various categories outperforms the framework using the ConvNets pre-trained on the small-scale dataset including only "Movies and TV" category which is semantically similar topic to the target dataset.

Table 17. The results of experiments for transfer learning on the IMDb review dataset. Each of numbers indicates the classification accuracy. (Various & Large) implies that each of models is pre-trained on the large-scale pre-training dataset including various categories (Table 13). (Movies & Small) implies that each of models is pre-trained on 150,000 Amazon reviews including only "Movies and TV" category [20].

Model	Accuracy
Bag-of-Words	0.81184
Bag-of-N-grams	0.83276
Scratch	0.75288
Pre-trained feature (Various & Large)	**0.87408**
Fine-tuning (Various & Large)	0.87396
Pre-trained feature (Movies & Small)	0.85156
Fine-tuning (Movies & Small)	0.85256

For the experiment for the Japanese dataset, we construct the large-scale dataset for pre-training from the Rakuten market review dataset. Table 18 shows an overview of the pre-training dataset.

We obtained a Rakuten travel review dataset from the Informatics Research Data Repository of National Institute of Informatics[8]. The Rakuten travel review

---

[7] http://www.imdb.com/.

[8] http://www.nii.ac.jp/dsc/idr/en/rakuten/rakuten.html.

Table 18. Overview of Rakuten market review dataset for pre-training.

Overall	Polarity	Train	Validation	Test
★5	Positive	300,000	20,000	20,000
★4		150,000	10,000	10,000
★2	Negative	150,000	10,000	10,000
★1		300,000	20,000	20,000

dataset has a format almost similar to the Rakuten market review dataset. We construct the small-scale dataset for the target task from the Rakuten travel review dataset. Table 19 shows an overview of the target dataset. The ratio of the number of the reviews of the target dataset for each of ★1, ★2, ★4 and ★5 rating is 2:1:1:2. The test classification accuracy of Small-C6FC3 one-hot for the pre-training dataset is 0.9382 and test classification accuracy of Small-C6FC3 embedding for the pre-training dataset is 0.9459.

Table 19. Overview of the Japanese dataset for transfer learning.

	Rakuten travel review dataset
Train	30,000
Validation	15,000
Test	15,000

Table 20 shows the results. The results indicate that the embedding model with Fine-tuning is the best method for all the datasets. From these facts, transfer learning of the character-level ConvNets is very effective also in Japanese dataset.

Table 20. The results of experiments for transfer learning on the Rakuten market review dataset and the Rakuten travel review dataset. Each of numbers indicates the classification accuracy.

Model		Accuracy
Bag-of-Words		0.9302
Bag-of-N-grams		0.9208
Scratch	one-hot	0.9151
Pre-trained feature	one-hot	0.9159
Fine-tuning	one-hot	0.9291
Scratch	embedding	0.9255
Pre-trained feature	embedding	0.9295
Fine-tuning	embedding	**0.9395**

5   Discussions

5.1   Features Extracted by Character-Level ConvNet

Figure 1 is a visualization of one filter weight matrix in the first layer of the Small-C6FC3 trained on the Amazon review dataset. Vertical direction of the visualization corresponds to window size, while horizontal direction of that corresponds to the dimensionality of one-hot encoding. In the visualization, white indicates large positive values, and black indicates large negative values, and gray indicates values close to zero. The weight matrix corresponding to letters from "a" to "z" has large variances for the vertical direction comparing with the weight matrix corresponding to other characters. This fact indicates that the ConvNet has learned to care more about the variations in letters than other characters. This phenomenon is observed in other filters as shown in Fig. 2. The visualization is consistent with one shown by Zhang et al.

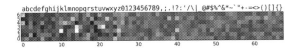

Fig. 1. Visualization of one filter weight matrix in the first layer of the Small-C6FC3 trained on the Amazon review dataset [20].

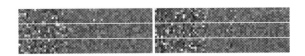

Fig. 2. Visualization of random sampled filter weights in the first layer of the Small-C6FC3 trained on the Amazon review dataset [20].

The visualization of filters from Small-C1FC1 trained on AFPBB dataset, arranged like Fig. 2, is shown in Fig. 3. The phenomenon is not observed in the filters of Small-C1FC1 trained on the AFPBB dataset. We suppose that these filters could not extract general features because of the smallness of the AFPBB dataset.

The visualization of the filter weight matrix of the ConvNet is shown in the past study. It is, however, hard to understand what the ConvNet extracts from the input text. Therefore, we attempt to investigate N-gram features to which a convolution filter of the ConvNet strongly responds. Since the window size of the first layer of all the ConvNets in this study is set to 7, we measure the output value of a convolution filter for each of 7-g and make a ranking based on the value as shown in Table 21. The filter is trained on the AG news dataset. In order to qualitatively evaluate the result, we extract articles which have these 7-g strings

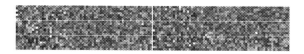

Fig. 3. Visualization of random sampled filter weights in the first layer of the Small-C1FC1 trained on the AFPBB dataset [20].

Table 21. Ranking of the output values of the convolution of one filter and 7-g. The filter is trained on the AG news dataset [20].

7-g		Convolution output value
ave	$70	1.02668
ove	$70	0.98627
ave	$10	0.97948
ts:	$10	0.96571
ise	$10	0.95659
the	$19	0.95559
9;s	$10	0.94505
ove	$10	0.93909
ave	$72	0.93072
ave	$20	0.93069

as its substring. We find that many of these articles have "save (or above or raise) $(the number)" as its substring. Actually, many of these articles are related to corporate cost-cutting, a sharp rise in oil prices, etc. Furthermore, for the quantitative evaluation, we extract articles which match the regular expression corresponds to the disjunction of all the 7-g shown in Table 21 and count the number of the matched articles belong to each category. The result is 22 for "Business", 0 for "Sports", 1 for "World" and 3 for "Sci/Tech". The result indicates that the feature extraction by the character-level ConvNets works well for text understanding. We suppose that the filter is able to represent at least $5 \times 4 \times 3 \times 1 \times 1 \times 3 \times 3 = 540$ kinds of 7-g.

In order to see if the effect mentioned above is language-dependent or not, we investigate one more filter which is trained on the romanized AFPBB dataset. Table 22 indicates the ranking of the output values of the convolution filter for 7-g. We find that the 7-g often occur in articles in the form of "(the one digit number)-(the one digit number) degyakuten/dehikiwake". The above romanized Japanese text means a come-from-behind win/a drawn match by a score of (the one digit number) to (the one digit number). Actually, many of these articles are related to sports such as tennis. If a classifier uses only the feature "degyakuten" (which means a come-from-behind win), the classifier could not distinguish sports articles from political articles since the feature "degyakuten" could be included in the political articles about elections. This result indicates that the feature extraction works well also in Japanese dataset.

Table 22. Ranking of the output values of the convolution of one filter and 7-g. The filter is trained on the romanized AFPBB dataset.

7-g	Convolution output value
,1-2deg	0.50774
a3-2deg	0.50613
a1-1deh	0.49199
a2-2deg	0.48423
a3-2deb	0.48332
a3-2def	0.47848
a1-3deh	0.47644
a3-1deh	0.47461
a2-1deg	0.47368
,3-2deh	0.46938

5.2   Transfer Learning in Text Classification

C6FC3 models have a large number of parameters. It is hard for the deep ConvNets to learn appropriate parameters which maximize the generalization ability if the number of the samples of the training dataset is small. Since all the datasets in the experiments in Sect. 4.3 are relatively small-scale, the Scratch frameworks, in other words, the vanilla deep ConvNets are almost the same or less accurate than the Bag-of-Words model and the Bag-of-N-grams model. The transfer learning frameworks, however, outperform the Bag-of-Words model and the Bag-of-N-grams model.

In addition, the experimental results on the IMDb review dataset imply that scale of the pre-training dataset is more important than a topic similarity between the pre-training dataset and the target dataset.

6   Conclusion

This study improves the classification accuracy by applying the character-level ConvNets to a large-scale Japanese text corpus in comparison with classic text classification methods. We analyze the features extracted by the character-level ConvNets. The result of the analysis shows that one of the filters of the convolutional layer of the ConvNet could represent multiple N-grams. We confirmed that meaningful representations are extracted by the ConvNets in English corpus and Japanese corpus. We revealed that the character-level ConvNets could eliminate language dependent preprocessing like the syntactic analysis. In addition, we provide the possibility of transfer learning by the ConvNets for text classification. We reuse the ConvNets pre-trained on the large-scale dataset to initialize the weights of the ConvNets for a target task which consists of relatively small dataset. This transfer learning framework improves the generalization ability and prevents from overfitting for the target task just like in the field of image recognition.

As future work, we would like to investigate transfer learning from a task to another task (*e.g.*, from a categorization task to a sentiment analysis task.). Additionally, we would like to pre-train the Japanese character-level embedding layer with a character-level skip-gram model or a Continuous Bag-of-Characters model inspired by the Continuous Bag-of-Words (CBoW) model.

Acknowledgements. This work was supported by JSPS KAKENHI Grant Numbers 26330081, 26870201, 16K12411, 17H04705. We use the Rakuten dataset which is provided by the National Institute of Informatics (NII) according to the contract between NII and Rakuten, Inc. We would like to thank NII and Rakuten, Inc.

References

1. Agrawal, P., Girshick, R., Malik, J.: Analyzing the performance of multilayer neural networks for object recognition. In: Fleet, D., Pajdla, T., Schiele, B., Tuytelaars, T. (eds.) ECCV 2014. LNCS, vol. 8695, pp. 329–344. Springer, Cham (2014). https://doi.org/10.1007/978-3-319-10584-0_22
2. Bengio, Y., Boulanger-Lewandowski, N., Pascanu, R.: Advances in optimizing recurrent networks. In: The Proceedings of the 2013 IEEE International Conference on Acoustics, Speech and Signal Processing (ICASSP 2013) (2013)
3. Del Corso, G.M., Gullí, A., Romani, F.: Ranking a stream of news. In: The Proceedings of the 14th International Conference on World Wide Web (WWW 2005), pp. 97–106 (2005)
4. Deng, J., Dong, W., Socher, R., Li, L.J., Li, K., Fei-Fei, L.: ImageNet: a large-scale hierarchical image database. In: The Proceedings of the 2009 IEEE Conference on Computer Vision and Pattern Recognition (CVPR 2009) (2009)
5. Girshick, R., Donahue, J., Darrell, T., Malik, J.: Rich feature hierarchies for accurate object detection and semantic segmentation. In: The Proceedings of the 2014 IEEE Conference on Computer Vision and Pattern Recognition (CVPR 2014) (2014)
6. Glorot, X., Bengio, Y.: Understanding the difficulty of training deep feedforward neural networks. In: The Proceedings of the 13rd International Conference on Artificial Intelligence and Statistics (AISTATS 2010) (2010)
7. Glorot, X., Bordes, A., Bengio, Y.: Domain adaptation for large-scale sentiment classification: a deep learning approach. In: The Proceedings of the 28th International Conference on Machine Learning (ICML 2011) (2011)
8. Gulli, A.: The anatomy of a news search engine. In: International Conference on World Wide Web (WWW) Special Interest Tracks and Posters, WWW 2005, pp. 880–881 (2005)
9. Kim, Y.: Convolutional neural networks for sentence classification. In: The Proceedings of the 2014 Conference on Empirical Methods in Natural Language Processing (EMNLP 2014), pp. 1746–1751 (2014)
10. Krizhevsky, A., Sutskever, I., Hinton, G.E.: Imagenet classification with deep convolutional neural networks. In: The Proceedings of the 26th Annual Conference on Neural Information Processing Systems (NIPS 2012), pp. 1097–1105 (2012)

11. Kudo, T., Yamamoto, K., Matsumoto, Y.: Applying conditional random fields to japanese morphological analysis. In: The Proceedings of the 2004 Conference on Empirical Methods in Natural Language Processing (EMNLP 2004), pp. 230–237 (2004)
12. Maas, A.L., Daly, R.E., Pham, P.T., Huang, D., Ng, A.Y., Potts, C.: Learning word vectors for sentiment analysis. In: The Proceedings of the 49th Annual Meeting of the Association for Computational Linguistics: Human Language Technologies (ACL HLT 2011), pp. 142–150 (2011)
13. McAuley, J., Pandey, R., Leskovec, J.: Inferring networks of substitutable and complementary products. In: The Proceedings of the 21th ACM SIGKDD International Conference on Knowledge Discovery and Data Mining (KDD 2015), pp. 785–794 (2015)
14. McAuley, J., Targett, C., Shi, Q., van den Hengel, A.: Image-based recommendations on styles and substitutes. In: The Proceedings of the 38th International ACM SIGIR Conference on Research and Development in Information Retrieval (SIGIR 2015), pp. 43–52 (2015)
15. Mikolov, T., Sutskever, I., Chen, K., Corrado, G.S., Dean, J.: Distributed representations of words and phrases and their compositionality. In: The Proceedings of the 27th Annual Conference on Neural Information Processing Systems (NIPS 2013), pp. 3111–3119 (2013)
16. Mikolov, T., Yih, W.T., Zweig, G.: Linguistic regularities in continuous space word representations. In: The Proceedings of the 2013 Conference of the North American Chapter of the Association for Computational Linguistics: Human Language Technologies (NAACL HLT 2013), pp. 746–751 (2013)
17. Nair, V., Hinton, G.E.: Rectified linear units improve restricted boltzmann machines. In: The Proceedings of the 27th International Conference on Machine Learning (ICML 2010), pp. 807–814 (2010)
18. dos Santos, C., Gatti, M.: Deep convolutional neural networks for sentiment analysis of short texts. In: The Proceedings of the 25th International Conference on Computational Linguistics (COLING 2014), pp. 69–78 (2014)
19. dos Santos, C.N., Xiang, B., Zhou, B.: Classifying relations by ranking with convolutional neural networks. In: The Proceedings of the 53rd Annual Meeting of the Association for Computational Linguistics (ACL 2015), pp. 626–634 (2015)
20. Sato, M., Orihara, R., Sei, Y., Tahara, Y., Ohsuga, A.: Japanese text classification by character-level deep ConvNets and transfer learning. In: The Proceedings of the 9th International Conference on Agents and Artificial Intelligence, vol. 2, pp. 175–184 (2017)
21. Severyn, A., Moschitti, A.: Twitter sentiment analysis with deep convolutional neural networks. In: The Proceedings of the 38th International ACM SIGIR Conference on Research and Development in Information Retrieval (SIGIR 2015), pp. 959–962 (2015)
22. Severyn, A., Moschitti, A.: UNITN: training deep convolutional neural network for Twitter sentiment classification. In: The Proceedings of the 9th International Workshop on Semantic Evaluation (SemEval 2015), pp. 464–469 (2015)
23. Sharif Razavian, A., Azizpour, H., Sullivan, J., Carlsson, S.: CNN features off-the-shelf: an astounding baseline for recognition. In: IEEE Conference on Computer Vision and Pattern Recognition (CVPR) Workshops, CVPR 2014 (2014)
24. Socher, R., Perelygin, A., Wu, J., Chuang, J., Manning, C.D., Ng, A.Y., Potts, C.: Recursive deep models for semantic compositionality over a sentiment treebank. In: The Proceedings of the 2014 Conference on Empirical Methods in Natural Language Processing (EMNLP 2013), pp. 1631–1642 (2013)

25. Srivastava, N., Hinton, G., Krizhevsky, A., Sutskever, I., Salakhutdinov, R.: Dropout: a simple way to prevent neural networks from overfitting. J. Mach. Learn. Res. **15**, 1929–1958 (2014)
26. Zhang, X., LeCun, Y.: Text understanding from scratch. CoRR abs/1502.01710 (2015)
27. Zhang, X., Zhao, J., LeCun, Y.: Character-level convolutional networks for text classification. In: The Proceedings of the 29th Annual Conference on Neural Information Processing Systems (NIPS 2015), pp. 649–657 (2015)

A Hierarchical Playscript Representation of Distributed Words for Effective Semantic Clustering and Search

Avi Bleiweiss[✉]

BShalem Research, Sunnyvale, USA
avibleiweiss@bshalem.onmicrosoft.com

Abstract. Semantic word embeddings have shown to cluster in space based on linguistic similarities that are quantifiably captured using simple vector algebra. Recently, methods for learning distributed word vectors have progressively empowered neural language models to compute compositional vector representations for phrases of variable length. However, they remain limited in expressing more generic relatedness between instances of a larger and non-uniform sized body-of-text. A recent study proposed a formulation that combines a word vector set of variable cardinality to represent a verse, with an iterative distance metric to evaluate similarity in pairs of non-conforming verse matrices. In this work, we expand on this sentence abstraction and apply it to a dialogue text passage that is prescribed in a playscript and uttered by an actor. In contrast to baselines characterized by a bag of features, our model preserves word order and is more sustainable in performing semantic matching at any of a dialogue, act, and play levels. To validate our framework for training word vectors, we analyzed the clustering of the complete play set of Shakespeare by exploring multidimensional scaling for visualization, and experimented with playscript searches of both contiguous and out-of-order parts of dialogues. We report robust results that support our intuition for measuring play-to-play and dialogue-to-play similarity.

1 Introduction

The attraction of using vector spaces for analyzing natural language semantics, stems primarily from providing an instinctive relation mechanism by subscribing to lexical distance and similarity concepts. In a large corpora of text, the structure of a semantic space is created by evaluating the various contexts in which words occur. Thus leading to distributional models of word meanings with the underlying assertion that words who transpire in similar contexts tend to have similar interpretations [26]. Distributed words, also known as word embeddings, are each represented with a dense, low-dimensional real-valued vector, and follow efficient similarity calculations directly from the known Vector Space Model [22]. Word vectors have been widely used as features in a diverse set of computational linguistic tasks, including information retrieval (IR) [17], parsing [24], named entity recognition [10], and question answering [12].

© Springer International Publishing AG, part of Springer Nature 2018
J. van den Herik et al. (Eds.): ICAART 2017, LNAI 10839, pp. 82–101, 2018.
https://doi.org/10.1007/978-3-319-93581-2_5

In recent years, neural network architectures have inspired the deep learning of word embeddings from large vocabularies to avoid a manual labor-intensive design of features. The work by Bengio *et al.* [4] introduced a statistical language model formulated by the conditional probability of the next word given all its preceding words in a sequence, or a context. However, the time complexity of the neural model renders the scheme inefficient due to the non-linear hidden layer. The succeeding Word2Vec [18] and GloVe [20] methods preserve the probabilistic hypotheses founded in Bengio *et al.* [4] approach, and further eliminate the hyperbolic tangent layer entirely, thus becoming more effective and feasible tools for language analysis. Notably, these methods expand on the input context window to include the words that both precede and follow the target word. Word embeddings are typically constructed by way of minimizing the distance between words of similar contexts, and encode various simple lexical relations as offsets in vector space. Our work investigates the linguistic structure in raw text, and explores clustering and search tasks using Word2Vec to train the underlying word representations.

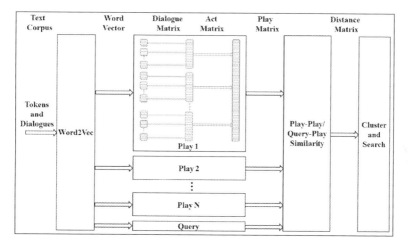

Fig. 1. Framework overview: on the left, tokens and context from a text corpus are used to train word vectors. A collection of word vectors is constructed to represent word-for-word the source text of every playscript. Word vectors are first row bound into a matrix to represent an actor specific or simultaneous dialogue. Then, dialogue matrices are concatenated into act matrices that are further coalesced into a hierarchical play matrix. We run all-play-pairs and all-query-play-pairs similarity process on matrices of a non-uniform row count, and generate a distance matrix that we use for cluster analysis and search ranking, respectively.

Applying unsupervised learning [8] of distributed word embeddings to a broader set of semantic tasks has not yet been fully established and remains an active research to date. In their recent work, Fu *et al.* (2014) utilize word embeddings to discover hypernym-hyponym type of linguistic relations. Noting that offsets of word pairs distribute into structured clusters, they modeled fine-grained relations by estimating projection matrices that map words to their respective hypernyms, and report a reasonable test F1-score of 73.74%. Socher *et al.* (2013) proposed a recursive neural network to compute distributed vector representations for phrases and sentences of variable length. Their model outperforms state-of-the-art baselines on both sentiment classification and accuracy metrics, however, its supervised method requires extensive manual labeling and makes scaling to larger sized text non trivial. A representation more rooted in a convolutional neural architecture [15], produces a feature map for each possible window in a sentence, and follows with a max-over-time pooling [6] to capture the most important features. Pooling has the apparent benefit of naturally adapting to variable length sentences. At the higher document level, Le and Mikolov [16] introduced a paragraph vector extension to the learning framework of word vectors. Given their different dimensionality, paragraph and word vectors are concatenated to yield fixed sized features, however, unique paragraph vectors constrain context sharing across the document.

For a composition of words, most of the techniques discussed tend to reshape the varying dimensionality of input sentences into uniform feature vectors. Rather, our implementation retains the word vectors as distinct rows in a matrix form to construct any of the dialogue, act, or play data structures for performing our set of linguistic tasks over a collection of scripts. The main contribution of our work is a framework (Fig. 1) with a lexical representation that abides word-for-word by the corpus source sequence, to facilitate a generic evaluation of relationships among entities of non-uniform text size. This work extends a recent study [5] by issuing responses to keyphrase queries that not only identify the act and scene enumerations inside a script, but in addition single out the corresponding actor names paired with the enclosed dialogue text-sequence. The rest of this paper is organized as follows. In Sect. 2, we briefly review the neural models to found Word2Vec, and proceed with motivating our choice for a dialogue matrix representation that leads to our act and play hierarchies. Section 3 derives our play similarity measure as it applies to a pair of non-conforming concatenations of dialogue embeddings, whereas Sect. 4 profiles the compiled format of the Shakespeare play-set corpus we used for evaluation. We then present our methodology for analyzing clusters of Shakespeare genres and ranking playscript searches of unsolicited keywords, and report extensive quantitative results of our experiments, in Sect. 5. We conclude with a discussion and future prospect remarks in Sect. 6.

2 Embedding Hierarchy

In Word2Vec, Mikolov *et al.* (2013a) proposed a shallow neural-network structure for learning useful word embeddings to support predictions within a local bi-directional context-window. Both the skip-gram and continuous bag-of-words (CBOW) models offer a simple single-layer architecture based on the inner product between a pair of word vectors. In the skip-gram version the objective is to predict the not necessarily immediate context words given the target word, and conversely, CBOW estimates the target word based on its neighboring context. As a context window scans over the corpus, the models attempt to maximize the log probability of the generated objective function, based on their respective multiple and single output layers, and training word vectors proceeds in a stochastic manner using back propagation. To improve upon accuracy and training time, Mikolov *et al.* (2013b) introduced both randomly discarding of frequent words that exceed a prescribed count threshold, and the concept of negative sampling that measures how well a word pairs with its context drawn from a noise distribution of a small sample of tokens. Empirically, neural model performance shown mainly governed by tunable parameters, including the word vector dimension, d, the symmetric context-window size, c_w, and the number of negative samples, s_n. Overall, skip-gram works well with a small amount of training data, while CBOW is several times faster to train.

The corpus we used for our study comprises a set of tens of playscripts and to train word embeddings, we first flattened the entire corpus into a linear array of dialogue text passages. We then proceeded to construct our basic data structures that culminate in an effective hierarchical representation of a play object, which is perceived nameless across subsequent clustering and search analyses. Let $w^{(k)} \in \mathbb{R}^d$ be the d-dimensional word vector corresponding to the k-th word in a dialogue. A dialogue text sequence S of length n is represented as a matrix

$$S = w^{(1)} \oplus w^{(2)} \oplus \ldots \oplus w^{(n)}, \tag{1}$$

where $\oplus$ is a row-wise binding operator and $S \in \mathbb{R}^{n \times d}$. S is thus regarded as a vector set and rows of S are considered atomic. To index a word vector in a dialogue, we use the notation s_k. Similarly, a play act A of m dialogues becomes a concatenation of dialogue matrices, $A = S^{(1)} \oplus S^{(2)} \ldots \oplus S^{(m)}$, where $A \in \mathbb{R}^{r_a \times d}$ and $r_a = \sum_{j=1}^{m} |S^{(j)}|$, and a play P comprises l act matrices, $P = A^{(1)} \oplus A^{(2)} \ldots \oplus A^{(l)}$, where $P \in \mathbb{R}^{r_p \times d}$ and $r_p = \sum_{i=1}^{l} r_a^{(i)}$. Respectively, a_j itemizes a dialogue matrix in an act, and p_i enumerates an act matrix in a play. Equation 2 provides an alternate matrix notation for each a dialogue, act, and play constructs.

$$S = \begin{bmatrix} w^{(1)} \\ w^{(2)} \\ \vdots \\ w^{(n)} \end{bmatrix} \qquad A = \begin{bmatrix} S^{(1)} \\ S^{(2)} \\ \vdots \\ S^{(m)} \end{bmatrix} \qquad P = \begin{bmatrix} A^{(1)} \\ A^{(2)} \\ \vdots \\ A^{(l)} \end{bmatrix}. \tag{2}$$

The length of a dialogue, $|S^{(j)}|$, and the number of dialogues per act, $|A^{(i)}|$, are varying parameters that we track and keep in a play map for traversing the play hierarchy. For play similarity computations, we often iterate a play matrix and access the entire collection of word vectors. Conveniently, we use a three dimensional indexing scheme, p_{ijk}, where each of i, j, and k points to an act matrix, dialogue matrix, and word vector, respectively. Space complexity for play embeddings is linear, $\mathcal{O}(lmn)$, and for a vocabulary that permits storing a 16-bit token enumeration instead, memory area required is thus reduced by a half. We further denote the corpus play set $T = \{P^{(1)}, P^{(2)}, \ldots, P^{(N)}\}$, where N, or cardinality $|T|$, is the number of plays.

3   Play Similarity

Measuring similarity and relatedness between distributed terms is an important problem in lexical semantics [1]. Recently, the process of learning word embeddings transpired compelling linguistic regularities by simply probing the linear difference between pairs of word vectors. This evaluation scheme exposes relations that are adequately distributed in a multi-clustering representation [9]. However, a single offset term is insufficient to assess similarity between a pair of plays represented by non-conforming matrices, each potentially retaining many thousands of word vectors. For evaluating semantic closeness of a pair of plays, $p^{(u)}$ and $p^{(v)}$, we explored a similarity concept that expands on the Chebychev matrix distance [5] and is defined by

$$ \mathrm{d}(u, v) = \frac{1}{|p^{(u)}|} \sum_{xyz} \left\{ \max_{ijk} \left(\mathrm{sim}(p^{(u)}_{xyz}, p^{(v)}_{ijk}) \right) \right\}, \tag{3} $$

where $|p^{(u)}|$ is the play cardinality that amounts to the total number of distributed word vectors for representing the play, and $|p^{(u)}| \neq |p^{(v)}|$. Whereas $\mathrm{sim}()$ is a similarity function that operates on two word vectors and takes either a Euclidean or an angle form. We chose cosine similarity [2] that performs an inner product on a pair of normalized vectors g and h, $\frac{g \cdot h^T}{\|g\|_2 \|h\|_2}$, and returns a scalar value as a measure of proximity. The time complexity of the distance algorithm is roughly quadratic, as for each word vector in play $p^{(u)}$, we find the closest word vector in play $p^{(v)}$, and then calculate the mean of all the closest distances derived formerly.

   In our system, all possible pairs of the corpus playscripts, T, are evaluated for similarity in the context of a $|T|$-dimensional squared distance-matrix, D. Elements of the distance matrix are considered directional and hence imply $\mathrm{d}(u, v) \neq \mathrm{d}(v, u)$. Matrix D facilitates unsupervised learning for clustering plays that apart from knowing their individual representations, they are perceived as a collection of unlabeled objects. Following an identical vein, as the distance metric provided by Eq. 3 is generic and assumes opaque matrix pairs, our framework naturally extends the semantic distance intuition to express a query-play type of relations for conducting a search. A query, q, comprises an unsolicited keyphrase and as such abides by our dialogue matrix formulation, S. The distance $\mathrm{d}(q, u)$, where $u \in \{1, 2, \ldots, |T|\}$, thus conveys a distinct relevancy measure for ranking the search of a query in each of the corpus plays contained in collection T. Our system reports back search results as a table of sorted distances paired with the play enumeration [7].

4   Shakespeare Corpus

We evaluated the performance of our model on playscripts written by William Shakespeare that we acquired online from the publicly available repository provided by Project Gutenberg [23]. The project cites the complete work of Shakespeare and comprises both plays and poems. Among the editions offered, we chose the plain-text eBook version that is distributed in a single file with instructive header separators between plays. Plays written by William Shakespeare are often divided into three major genres and include comedy, history, and tragedy type of content, as Table 1 lists the play names associated with each of the genre partitions. Our study pertains to mining text of a traditional playscript format, and thus excludes poetry type compositions by Shakespeare, such as the renown collection of 154 sonnets.

Table 1. List of Shakespeare play names classified by their respective genres.

Comedy	History	Tragedy
All's Well That Ends Well	Henry IV, part 1	Antony and Cleopatra
As You Like It	Henry IV, part 2	Coriolanus
The Comedy of Errors	Henry V	Hamlet
Cymbeline	Henry VI, part 1	Julius Caesar
Love's Labours Lost	Henry VI, part 2	King Lear
Measure for Measure	Henry VI, part 3	Macbeth
The Merry Wives of Windsor	Henry VIII	Othello
The Merchant of Venice	King John	Romeo and Juliet
A Midsummer Night's Dream	Richard II	Timon of Athens
Much Ado About Nothing	Richard III	Titus Andronicus
Pericles, Prince of Tyre		
Taming of the Shrew		
The Tempest		
Troilus and Cressida		
Twelfth Night		
Two Gentlemen of Verona		
Winter's Tale		

The Shakespeare dataset consists of 37 full-length titles, as 17, 10, and 10 plays subscribe to comedy, history, and tragedy genre groups, respectively. Every Shakespeare playscript contains a section header labeled dramatis personae, persons of the drama, that lists the main characters of the play. However, this list is often incomplete, as there are many background and offstage roles that emerge as the storytelling evolves. For our work, we have scanned each playscript from start to end and manually extracted a complete list of all the names of active actors in

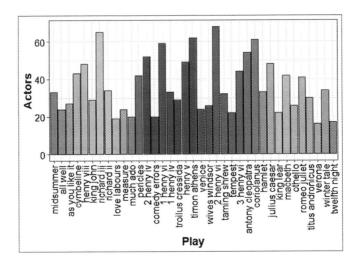

Fig. 2. Actor distribution across the entire Shakespeare play suite.

a play, and used it as a search index for binding a dialogue to an actor. In total, the Shakespeare corpus incorporates 1,352 distinct characters and Fig. 2 provides visualization of actor distribution across the entire play suite. Uniformly, every Shakespeare play is divided into five acts, as each act is composed of one or more scenes. For a scene, the playscript contains a brief header description of location and time of day, and may follow by appearance directions such as characters entering and exiting the stage. Successively, the script presents a chronology of dialogue text passages, each identified with a single or group of characters. A dialogue is often a short text sequence, but may also comprise a paragraph of several sentences. The Shakespeare corpus combines a total of 749 scenes and 30,473 dialogues, and visualization of per-act stacked distribution of scenes and dialogues across all the plays is outlined in Figs. 3 and 4, respectively. Table 2 provides complementary summarizations of actors-per-play, and aggregations over acts for scenes-per-play, and dialogues-per-play.

To derive our word vectors, we first tokenized and lowercased the dialogue text passages of the entire play suite, removed all punctuation marks excluding the hyphen, and have retained stop words of Old English that dates back to the Shakespeare era of the 17th century. The corpus has a little under two million unfiltered symbols with a vocabulary of 24,747 unique tokens to train (Table 2). Notably, most unsolicited tokens are of a low frequency term with only 22 symbols exceeding 100 incidents. The mean frequency term is 56, and there are 196 tokens, under one percentage point of the vocabulary size, that occur only once in the dataset. In Fig. 5, we show our word cloud rendition using R [21] that depicts the top 100 frequent tokens in a linear dialogue collection extracted from the Shakespeare playscript suit. As anticipated, words of relational connotations like 'thou', 'thy', and 'thee', or of a romantic association such as 'good' and 'love', and monarchy related 'lord' and 'king' are of the highest occurrence term.

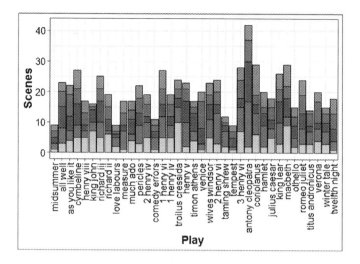

Fig. 3. Per-act stacked scene distribution across the entire Shakespeare play suite.

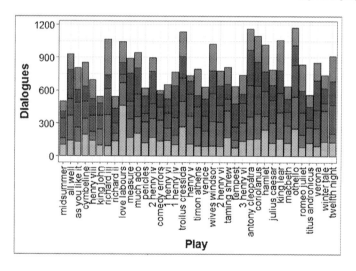

Fig. 4. Per-act stacked dialogue distribution across the entire Shakespeare play suite.

Dialogue text passages are regularly preceded in the playscript with a pattern consisting of an uppercased character name that is followed by a period. We note that this pattern remains verbatim and excluded from tokenization, and internally we retain a dialogue data structure as a record pair of the actor name and his or her corresponding text sequence. This lets us at any time respond to search queries with both a playscript location that comprises the act and scene IDs, and in addition provide the respective actor name. In addition, dialogues paired with multiple actors are fairly regular in Shakespeare plascripts, including the common appearances of LORDS, SERVANTS, ALL, and CHORUS, and both our actor index and representation abide by this simultaneous rule.

Fig. 5. caption A word cloud rendition of the top 100 frequent tokens as they appear in a linear dialogue collection extracted from the full set of Shakespeare playscripts. Font size is proportional to the number of word occurrences in the corpus.

To construct a context window, we randomly select an unlabeled play enumeration in the [1–37] range, and traverse our hierarchy top-to-bottom by arbitrarily sampling act and dialogue indices that are confined to the limits set by the singled out playscript. In our implementation, running dialogue indices cross scene boundaries and are presumed continuous throughout a given play act. In the text passage of the chosen dialogue, a random target-word position is used to extract from left and right context words that are delimited by the dialogue start and end words. Ultimately, one of our system goals is to discover semantic relations that closely align learned play clusters with the manually-prescribed classified genres listed in Table 1.

5   Empirical Evaluation

Previously, we discussed vector embedding techniques, such as Word2Vec [18] and GloVe [20], and their role to transform natural language words into a semantic vector space. In this section, we proceed to quantitatively evaluate the intrinsic quality of the produced set of latent vector representations, and analyze their impact on the performance of our unsupervised learning tasks that comprise play clustering and search. As an aid to tune our system level performance, we explored varying some of the hyperparameters designed to control the neural models incorporated in the word embedding methods. In practice, we have implemented our own Word2Vec technique natively in R [21] for better integration with our software framework. Across all of our experiments, we trained word vectors employing mini-batch stochastic gradient descent (SGD) with an annealed learning rate, and semantic similarity results we report on both play-to-play and dialogue-to-play relations presume anonymous plays.

Table 2. Summarizations of play properties including actors-per-play, aggregations over acts for scenes-per-play and dialogues-per-play, and at a corpus level the total of unsolicited and unique tokens.

Actors			
Min	Max	Mean	Total
16	168	36.5	1,352

Scenes			
Min	Max	Mean	Total
9	42	20.2	749

Dialogues			
Min	Max	Mean	Total
499	1,175	823.6	30,473

Tokens	
Unsolicited	Unique
1,999,034	24,747

5.1   Experimental Setup

The raw text of Shakespeare playscripts [23] underwent a data cleanup preprocess to decompose hyphenated word compounds, properly uppercase all instances of actor declarations, and introduce a more definite separation symbol between an actor ID and the succeeding dialogue text-sequence. We then tokenized the corpus using the R tokenizer [21] and built a maximal vocabulary V of size $|V| = 24,747$. Each word in this sparse 1-of-$|V|$ encoding is represented as a one-hot vector $\in \mathbb{R}^{|V| \times 1}$, with all its components zeroed out and a single one set at the index of the word in the vocabulary, that is further mapped onto a lower-dimensional semantic vector-space. Projecting onto the denser formulation transpires at the stage immediately preceding the hidden layer of the neural models that underlie the embedding technique.

In the absence of a large supervised training set of word vectors, the use of pre-trained word vectors obtained from an unsupervised neural model became a favorable practice to boost system performance [6,12,15]. While proven useful for word analogy and multi-class classification tasks, clustering and search over a rather unique dataset requires however randomly initialized word-vectors. Hence our model generates distinct input and output sets of word vectors, W and $\tilde{W}$, that only differ as a result of their random initialization. To help reduce overfitting and noise, we used their sum, $W + \tilde{W}$, as our final vectors and that

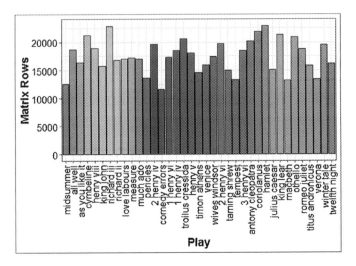

Fig. 6. Flattened play hierarchy into a single linear matrix of word vectors $w^{(k)} \in \mathbb{R}^d$. Showing distribution of matrix row count across the entire Shakespeare play suite.

typically yields a small performance gain. To better assess the space complexity of our play representation made of the trained word embeddings, Fig. 6 provides a visual interpretation of a flattened play hierarchy, outlined as a single linear matrix with up to several tens-of-thousands rows of word vectors, and shown distributed across the entire Shakespeare play suite.

Recent study by Baroni *et al.* [3] alluded to neural word-embedding models that consistently outperform the more traditional count-based distributional methods on many semantic matching tasks and by a fair margin. Furthermore, much of the achieved performance gains cited are mostly attributed to a system design choice of the configured hyperparameters. Motivated by these results, we evaluated task performance comparing distinct play hierarchies generated by each skip-gram and CBOW, and choose negative sampling that typically works better than hierarchical softmax [19]. For the learning hyperparameters, there seems no single selection for an optimal word-vector dimension, d, as it tends to be highly task dependent and varies from 25 for semantic classification [24] up to 300 for word analogy [18]. Rather, we set $d = 10$ and traded-off word vector dimension to attain more tractable computation in building the distance matrix that is inherently of a quadratic time complexity, $\mathcal{O}(|T|^2)$. Whereas, to better assess the impact of the context window on our system performance, we varied discretely its size $c_w = \{5, 15, 25, 50\}$, in a wide range of word counts. Evidently, Word2Vec performance tends to decrease as the number of negative samples increases beyond about ten [20], thereby we used $s_n = 10$. For our mini-batch SGD to train word vectors, we used a batch size of 25 and an initial learning rate $\alpha = 0.1$, and updated parameters after each window. The number of epochs we ran in our experiments topped at 10,000, and is determined by both the chosen neural model and context window size, c_w.

5.2   Experimental Results

We present play clustering results of our own trained Shakespeare corpus at both the genre component level and for the entire suite of $|T| = 37$ plays. To visualize genre based clusters, we used Multidimensional Scaling (MDS) [11,25] that extracts patterns of semantic proximities from our play distance-matrix representation, D. The distance matrix is composed of a set of pairwise play-similarity values with $\mathcal{O}(|T|^2)$ scaling that MDS further compiles and projects onto an embedding p-dimensional Euclidean-space. This mapping is intended to faithfully preserve the clustering structure of the original distance data-points, and often, data visualization quality of clusters is directly proportional to the ratio $\frac{p}{|T|}$. In our experiments, we consistently rendered similarity measures of play pairs onto a two-dimensional coordinate frame to inspect and analyze formations of genre-based play clustering.

In Fig. 7, we provide baseline visualization of MDS applied to our play distance matrices that represent each of the genre components of comedy, history, and tragedy with dimensionality $|T| = \{17, 10, 10\}$, respectively. For this experiment, we used the CBOW neural model to train word vectors, as hyperparameters were uniformly set to their defaults, using 5 words for the context window size, c_w. The comedy collection shows most of its plays semantically closely related, with the exception of A Midsummer Night's Dream as an outlier with a fair distance apart. Whereas the history genre class of ten play samples has eight of its members arranged as immediate neighbors, the position of Henry VIII slightly disjoint, and the first part of Henry IV noticeably apart. On the other hand, the tragedy genre selection formed two distinct adjacent clusters of five and three plays, respectively, as the titles Antony and Cleopatra and Timon of Athens are notably detached.

A much broader interest of our work underscores the cluster analysis of a single distance matrix with dimensionality $|T| = 37$ that represents the entire Shakespeare play suite. Through this evaluation, our main objective is to predict unsupervised play clustering and assess its matching to the Shakespeare formal

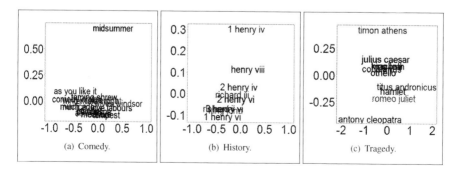

Fig. 7. Visualization of play distance matrices using Multidimensional Scaling. Representing the Shakespeare genres of comedy, history, and tragedy each with 17, 10, and 10 plays, respectively.

Table 3. Visualization of clustering the entire Shakespeare play suite using Multidimensional Scaling, as each play is assigned its formal genre-subset legend post projecting onto the displayable embedding space. Shown as a function of a non-descending context window-size, c_w, for both the skip-gram and CBOW neural models.

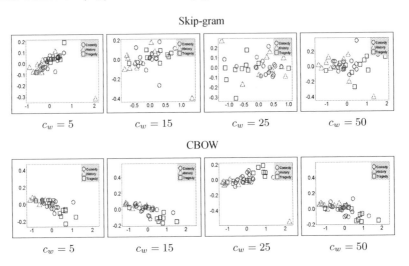

genre subdivisions (Table 1). Table 3 shows the clustering produced by applying MDS to the single matrix that captures all-play-pairs semantic similarities, as each play is assigned its formal genre-subset legend post projecting onto the displayable embedding space. In these experiments, we compared unique word-vector sets generated by each skip-gram and CBOW, as we stepped over the fairly large extent of discrete values prescribed for the context window size, c_w. Expanding the context window scope has a rather mild impact on cluster construction with word vectors trained by the CBOW neural model, noting that for $c_w = 25$, clusters are shown shifted upwards as the mapping onto the history genre gave rise to one outlier, the Henry V play, at the bottom right corner of the plot area. However for skip-gram, group formations are considerably susceptible and affected by even a moderate change of c_w. Furthermore, the play partitions we generated under CBOW training persistently resemble the official subdivision of Shakespeare genre collections, although for visualization in a 2D embedding space, the comedy and history sets do overlap each other. The impact of the neural model used on clustering playscripts largely concurs with the results obtained using an orthogonal bible book-suite [5].

In our play search experiments, we explored three types of keyphrase queries including fixed dialogue text-passages drawn from a known play and act, reordered words of random partial dialogues distributed uniformly in each of the plays of the Shakespeare suite, and randomly selected tokens from the corpus vocabulary composed into a set of keywords. Every search is preceded by converting the query composition into a matrix of word vectors and then pairing the query matrix with each of the play hierarchies to compute similarity distances. Thus resulting in a process of linear time complexity, $\mathcal{O}(|T|)$. Unless otherwise noted, for the search experiments we applied default hyperparameters to CBOW-based trained word-vectors.

Table 4. Search queries of fixed dialogue text-sequences from known play origins.

Play	Actor	Search Query
A Midsummer Night's Dream	HIPPOLYTA	*bent in heaven shall behold the night*
Much Ado About Nothing	HERO	*says the prince and my new trothed lord*
King John	BASTARD	*bragging horror so shall inferior eyes*
Hamlet	QUEEN	*did he receive you well*
Macbeth	MACBETH	*our will became the servant to defect*

In Table 4, we list search queries of fixed dialogue text-passages and correspondingly enumerate their play origins and bound actors. Overall, for each of the five dialogues searched the predicted play title matched the expected label and was ranked highest with a score of 1.0. King John appeared to be a single source play for its keyphrase as the rest of the plays scored fairly low with a mean of 0.27. However, not surprisingly the search of the remaining four queries uncovered additional unlisted plays that equally claimed a lead score, as keyphrase words either extend within an act or over acts non-adjacently and would still rank high for relevance in the context of a play search. For instance, the Twelfth Night play scored high on the keyphrase from the play of Hamlet with keywords split apart among the first three acts, and similarly Cymbeline ranked high for the dialogue text passage from Macbeth, as keyword groups appear in each of the first and fifth acts. Figure 8 provides visualization to our fixed-dialogue search results in the form of a depleted confusion matrix $\in \mathbb{R}^{37 \times 5}$, with actual and predicted plays listed at the bottom and to the left of the grid, respectively.

In the second search experiment, we selected from each of the Shakespeare plays five random samples of partial dialogues, each with eight unordered context words. We ran a total of $5 \times 37 = 185$ search episodes and constructed a search matrix by computing all directional pairs of query-play distances, and then averaged the score for multiple queries per play. In Fig. 9, we show our results for the random sub-dialogue search and demonstrate consistent top ranking for when the source play of the queries matches the predicted play along the diagonal of the fully populated confusion-matrix $\in \mathbb{R}^{37 \times 37}$.

Our third task deployed cumulatively a total of 200 searches using a query keyphrase that is a composite of randomly selected tokens from our entire vocabulary, and thus implies a weak contextual relation to any of the Shakespeare

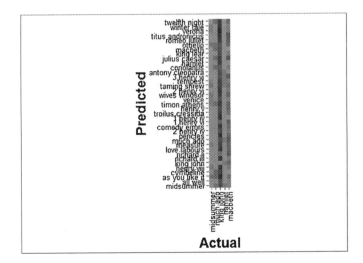

Fig. 8. Visualization of our fixed dialogue search results provided in the form of a depleted confusion matrix $\in \mathbb{R}^{37 \times 5}$.

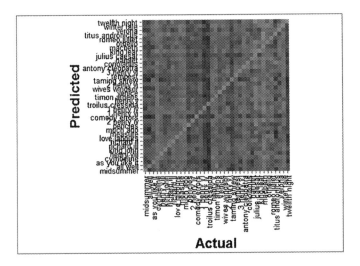

Fig. 9. Visualization of our randomly reordered sub-dialogue search results provided in the form of a fully populated confusion-matrix $\in \mathbb{R}^{37 \times 37}$.

plays. Distributed non uniformly, our token based keyphrases are evidently biased towards affiliation with plays of the largest content. Figure 10 shows non-zero query allocations that occupy 36 out of 37 plays, excluding the Comedy of Errors title. As a preprocess step, we iterated over the extended search matrix of dimensionality (37×200) and identified the play that is closest to a given query. We followed by averaging the distances in the case of multiple queries per play, and ended up with a reduced search matrix $\in \mathbb{R}^{37 \times 36}$. Figure 11 shows the

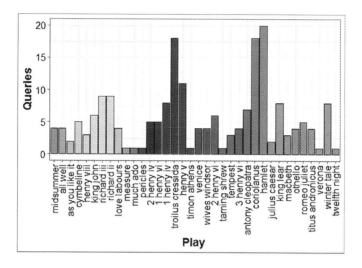

Fig. 10. Distribution of queries in random token search, presented across 36 Shakespeare titles. The larger the playscript content, the higher the query count.

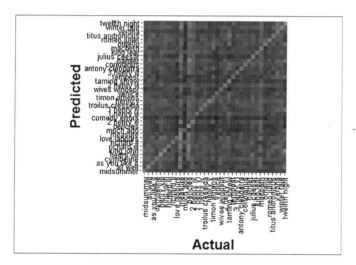

Fig. 11. Visualization of our random token search results provided in the form of a reduced dimensionality confusion-matrix $\in \mathbb{R}^{37 \times 36}$.

results of random token search as the straddling bright line along the diagonal of the confusion matrix highlights our best ranks. In this experiment, search scores are expected to be diverse and span a rather wide range of $[0.23, 0.87]$.

In Table 5, we list computational runtime of our implementation for key tasks in performing linguistic clustering and search. All reported figures are in seconds and were obtained by running our software single-threaded on a Windows 10 mobile device, with Intel 4th generation Core[TM] processor at 1.8 GHz and 8 GB

Table 5. Running time for key computational tasks in performing clustering and search over the Shakespeare corpus. All figures are shown in seconds.

Task	Min	Max	Mean	Total
Hierarchy Formation	0.58	1.44	0.91	33.53
Distance Matrix	0.83	80.08	56.19	2,079.35
Keyphrase Query	13.48	18.66	16.65	616.05

of memory. Play hierarchy construction is linear with the number of dialogues per play, and given a fairly balanced distribution over playscripts (Fig. 6), the play of Othello claimed the slowest to generate the data structure at 1.44 s. The distance matrix item shows the time to compute a set of similarities for one play paired with each of the rest of the plays in the Shakespeare collection. On average, play-to-play distance derivation amortized across $|T| = 37$ plays takes about 1.51 s. Launching a keyphrase query task typically involves a dialogue-to-play similarity operator that is linear in the total number of dialogues for the entire play collection. Query response times are shown for each search episode and are uniformly impartial to the keyphrase originating play, as evidenced by a small standard deviation of two percent of the mean. The total column of Table 5 further accumulates individual play processing times and is roughly the mean column value multiplied by the number of plays, $|T| = 37$.

6   Conclusions

In this study, we have demonstrated the apparent potential in a hierarchical representation of word embeddings to conduct effective play level clustering and search. We trained our system on a 37-dramatic-play corpus that comprises a vocabulary of about 2 million tokens, and generated our own word vectors for each of our experimental choices of model-hyperparameter configurations. We showed that the CBOW neural model outperforms skip-gram for the linguistic tasks we performed, and furthermore, clustering under CBOW proved sustainable to modifying the context window size in a fairly large extent. To evaluate any-pair semantic similarity of both play-to-play and query-to-play, we used a simple and generic distance metric [5] between a pair of word vector sets, each of up to tens of thousands elements, that disambiguates non-matching matrix dimensionality. We reported robust empirical results on our tasks for deploying state-of-the-art unsupervised learning of word representations.

At first observation, our hierarchical representation of plays might appear greedy storage wise, and rather than a matrix interpretation, we could have resorted to a more compact format by averaging all the dialogue word vectors and produce a single dialogue vector. While this approach seems plausible for the clustering tasks to both reduce footprint and streamline computations, the data loss incurred by doing so adversely impacted the performance of our search tasks. To address this shortcoming, our experimental choice of a modest 10-dimensional word vector appears as a reasonable system-design trade-off that aids to circumvent excessive usage of memory space. On average, there are 17,509 word vectors per-play (Fig. 6) and hence storage space for the entire Shakespeare play-suite is at a moderate $37 \times 17,509 \times 10 \times 4 \approx 26$ MB that is proportional to 34 MB area claimed by the bible book set [5].

To the best of our knowledge and based on literature published to date, we are unaware of semantic analysis systems with similar goals to evenhandedly contrast our results against. The more recent work by Yang *et al.* [27], proposed a hierarchical attention network for document classification. Their neural model explores attention mechanisms at both a word and sentence levels in an attempt to differentiate content importance when constructing a document vector representation. However, for evaluation their work focuses primarily on topic classification of short user-review snippets. Unlike our system that reasons semantic relatedness between any full-length plays. On the other hand, Jiang *et al.* [13] skip the sentence level construct altogether and combine a set of word vectors to directly represent a complete Yelp review. In their report, there is limited exposure to fine-grained control over the underlying neural models to show performance impact on business clustering.

Given that the training of word vectors is a one time process, a natural progression of our work is to optimize the core computations of constructing the distance matrix and performing a keyphrase query. The inherent independence of deriving similarity matrix elements and separating play search rankings lets us leverage parallel execution, and we expect to reduce our runtime complexity markedly. For a larger number of corpus plays, to uphold efficient hierarchy access our system would benefit from word vector caching, and we contend that projecting the distance matrix onto a three-dimensional embedding space is essential to improve cluster perception for analysis. Lastly, we seek to apply our play distance matrix directly to methods that partition objects around medoids [14] and potentially avoid outliers.

Acknowledgements. We would like to thank the anonymous reviewers for their insightful suggestions and feedback.

References

1. Agirre, E., Alfonseca, E., Hall, K., Kravalova, J., Paca, M., Soroa, A.: A study on similarity and relatedness using distributional and WordNet-based approaches. In: Human Language Technologies: North American Chapter of the Association for Computational Linguistics (NAACL), Stroudsburg, PA, pp. 19–27 (2009)
2. Baeza-Yates, R., Ribeiro-Neto, B. (eds.): Modern Information Retrieval. ACM Press Series/Addison Wesley, Essex (1999)
3. Baroni, M., Dinu, G., Kruszewski, G.: Don't count, predict! A systematic comparison of context-counting vs. context-predicting semantic vectors. In: Annual Meeting of the Association for Computational Linguistics (ACL), pp. 238–247 (2014)
4. Bengio, Y., Ducharme, R., Vincent, P., Janvin, C.: A neural probabilistic language model. Mach. Learn. Res. (JMLR) **3**, 1137–1155 (2003)

5. Bleiweiss, A.: A hierarchical book representation of word embeddings for effective semantic clustering and search. In: Agents and Artificial Intelligence (ICAART), pp. 154–163. Porto, Portugal (2017)
6. Collobert, R., Weston, J., Bottou, L., Karlen, M., Kavukcuoglu, K., Kuksa, P.: Natural language processing (almost) from scratch. Mach. Learn. Res. (JMLR) **12**, 2493–2537 (2011)
7. Cormen, T.H., Leiserson, C.H., Rivest, R.L., Stein, C.: Introduction to Algorithms. MIT Press/McGraw-Hill Book Company, Cambridge (1990)
8. Duda, R.O., Hart, P.E., Stork, D.G.: Unsupervised learning and clustering. In: Pattern Classification, pp. 517–601. Wiley, New York (2001)
9. Fu, R., Guo, J., Qin, B., Che, W., Wang, H., Liu, T.: Learning semantic hierarchies via word embeddings. In: Annual Meeting of the Association for Computational Linguistics (ACL), Baltimore, MD, pp. 1199–1209 (2014)
10. Guo, J., Che, W., Wang, H., Liu, T.: Revisiting embedding features for simple semi-supervised learning. In: Empirical Methods in Natural Language Processing (EMNLP), Doha, Qatar, pp. 110–120 (2014)
11. Hofmann, T., Buhmann, J.: Multidimensional scaling and data clustering. In: Advances in Neural Information Processing Systems, pp. 459–466. MIT Press, Cambridge (1995)
12. Iyyer, M., Boyd-Graber, J., Claudino, L., Socher, R., Daume, H.: A neural network for factoid question answering over paragraphs. In: Empirical Methods in Natural Language Processing (EMNLP), Doha, Qatar, pp. 633–644 (2014)
13. Jiang, R., Liu, Y., Xu, K.: A General Framework for Text Semantic Analysis and Clustering on Yelp Reviews (2015). http://cs229.stanford.edu/proj2015/003_report.pdf
14. Kaufman, L., Rousseeuw, P.J. (eds.): Finding Groups in Data: An Introduction to Cluster Analysis. Wiley, New York (1990)
15. Kim, Y.: Convolutional neural networks for sentence classification. In: Empirical Methods in Natural Language Processing (EMNLP), Doha, Qatar, pp. 1746–1751 (2014)
16. Le, Q.V., Mikolov, T.: Distributed representations of sentences and documents. CoRR abs/1405.4053 (2014). https://arxiv.org/abs/1405.4053
17. Manning, C.D., Raghavan, P., Schutze, H.: Introduction to Information Retrieval. Cambridge University Press, Cambridge (2008)
18. Mikolov, T., Chen, K., Corrado, G., Dean, J.: Efficient estimation of word representations in vector space. CoRR abs/1301.3781 (2013a). http://arxiv.org/abs/1301.3781
19. Mikolov, T., Sutskever, I., Chen, K., Corrado, G.S., Dean, J.: Distributed representations of words and phrases and their compositionality. In: Advances in Neural Information Processing Systems, pp. 3111–3119. Curran Associates Inc, Red Hook (2013b)
20. Pennington, J., Socher, R., Manning, C.D.: GloVe: Global vectors for word representation. In: Empirical Methods in Natural Language Processing (EMNLP), Doha, Qatar, pp. 1532–1543 (2014)
21. R Core Team: R: A Language and Environment for Statistical Computing. R Foundation for Statistical Computing, Vienna, Austria (2013). http://www.R-project.org/
22. Salton, G., Wong, A., Yang, C.S.: A vector space model for automatic indexing. Commun. ACM **18**(11), 613–620 (1975)
23. Shakespeare, W.: Project Gutenberg: The complete works of William Shakespeare (1994). http://www.gutenberg.org/ebooks/100

24. Socher, R., Perelygin, A., Wu, J.Y., Chuang, J., Manning, C.D., Ng, A.Y., Potts, C.: Recursive deep models for semantic compositionality over a sentiment treebank. In: Empirical Methods in Natural Language Processing (EMNLP), Seattle, WA, pp. 1631–1642 (2013)
25. Torgerson, W.S.: Theory and Methods of Scaling. Wiley, New York (1958)
26. Turney, P.D., Pantel, T.: From frequency to meaning: vector space models of semantics. Artif. Intell. Res. (JAIR) **37**, 141–188 (2010)
27. Yang, Z., Yang, D., Dyer, C., He, X., Smola, A., Hovy, E.: Hierarchical attention networks for document classification. In: Human Language Technologies: North American Chapter of the Association for Computational Linguistics, pp. 1480–1489 (2016)

Data Harvesting and Event Detection from Czech Twitter

Václav Rajtmajer[1] and Pavel Král[1,2(✉)]

[1] Department of Computer Science and Engineering,
Faculty of Applied Sciences, University of West Bohemia, Plzeň, Czech Republic
{rajtmajv,pkral}@kiv.zcu.cz
[2] NTIS - New Technologies for the Information Society,
Faculty of Applied Sciences, University of West Bohemia, Plzeň, Czech Republic
http://nlp.kiv.zcu.cz

Abstract. Twitter belongs to the fastest-growing microblogging and online social media. Automatically monitoring and analyzing this rich and continuous data stream can yield valuable information, which enable users and organizations to discover important knowledge. This paper proposes a method for harvesting of important messages from Czech Twitter with high download speed and an approach to discover automatically the events in such data. We identified important Twitter users and then we use these lists to discover potentially interesting tweets to download. The tweets are then clustered in order to discover the events. Final decision is based on the thresholding. We show that the harvesting method downloads about 6 times more data than the other approaches. We further report promising results of the event detection approach on a small corpus of the Czech Tweets.

Keywords: Czech · Clustering · Data · Event detection · Harvesting
Social media · Twitter

1 Introduction

Microblogging is a novel broadcast (social) medium that allows to create, share, view and analyze information particularly in a form of short messages. Although this service is relatively new compared to traditional media, it has gained significant popularity among individuals, companies and other organizations. For example, during recent social upheavals and crises, many people on the planet used Twitter to report and follow the main events. The importance and the size of the today's social media are growing very rapidly which is strictly related to the particular needs of the automatic processing methods.

Twitter is currently the most fastest-growing microblogging medium, with more than 250 million users producing about 500 million tweets[1] per day[2]. The

---

[1] Short messages limited by 140 characters.

[2] http://www.internetlivestats.com/twitter-statistics/ - June 2017.

© Springer International Publishing AG, part of Springer Nature 2018
J. van den Herik et al. (Eds.): ICAART 2017, LNAI 10839, pp. 102–115, 2018.
https://doi.org/10.1007/978-3-319-93581-2_6

tweets can be accompanied by photos, videos, geolocation, links to other users and trending topics. The posted tweet can be liked, commented by the other tweets, or redistributed by other users by forwarding, so-called *retweet*. Due to its simplicity and easy access, Twitter contains a wide range of topics from common every day conversations over sport news to information about ongoing disasters as earthquake, flood or typhoon.

Twitter is thus certainly an interesting source of real-time information which can be used for further analysis and data-mining. In this work, we use Twitter because of its large size, significant amount of other existing work dealing with this network and particularly because of a number of Twitter users post interesting news from various topics in real-time. We will use Twitter for automatic real-time event detection because it will be very useful for many journals and news agencies in order to discover quickly new interesting information. Particularly, the Czech News Agency (ČTK[3]) needs a system to automatically collect data from Czech Twitter and on-line discover potential events.

Our first task consists in analyzing Twitter stream and harvesting the appropriate Czech tweets in real-time. The second important task is the subsequent analysis of the downloaded data to discover new events. We have described in [1] a novel Twitter harvesting approach with high download speed. However the event detection was not evaluated precisely, because of the lack of an annotated corpus. This paper extend this work and first presents a small Czech Twitter corpus annotated with the events. Then, the event detection is evaluated on this corpus.

The core of the harvesting method relies on using user lists to download a sufficient number of Czech tweets in real-time. The tweets are then clustered to groups in order to discover the events. Final decision is based on the thresholding.

Several definitions of events exist, however we will use in this work the definition from Cambridge dictionary where an event is defined as "anything that happens, especially something important and unusual[4]".

The paper structure is as follows. Section 2 is gives a short review of Twitter analysis with a particular focus on event detection. Section 3 presents the proposed approach for Czech Twitter analysis and event detection. The following section describes the architecture of the proposed event detection system. Section 5 deals with the results of our experiments. In the last section, we conclude the paper and propose some future research directions.

2 Related Work

Twitter offers a number of possibilities for data processing and analysis, it is thus investigated by many recent research works. The data from this social medium can be used for instance for opinion mining or sentiment analysis as shown in [2]. In this paper, a corpus for sentiment analysis on Twitter is presented and an efficient sentiment classifier is further built on this data. Another work

---

[3] http://www.ctk.eu/.
[4] http://dictionary.cambridge.org/dictionary/british/event.

about sentiment analysis on Twitter is proposed in [3]. This paper deals with the importance of linguistic features for this task. The data from Twitter can also be used for sociological surveys as presented in [4]. This paper analyzes a group polarization using the data collected from dynamic debates. Another paper [5] deals with Twitter analysis to discover user activities. The authors present a taxonomy characterizing the underlying user intentions.

Twitter is also successfully used for event detection task as presented for instance in [6,7]. These approaches generally capture a presence or an increase of important key-words. A presence of words "hurricane" or "inundation" is used to detect the catastrophes.

It has been also suggested many sophisticated Twitter analysis methods as for instance in [8]. This paper describes a system called *Twevent*, which first detects the segments of events and then, they are clustered considering both their frequency distribution and content similarity for event detection. The authors use Wikipedia as a knowledge base to identify the most interesting segments to discover realistic events. The main advantage of this method from the previous ones is that it is domain independent and therefore, it can discover all types of the events.

Recently, bursty event detection techniques [9] have been emergent. These methods lie on the assumption that previously unseen or rapidly growing topics in the stream should represent new events. They thus consider an event in data streams as a bursty activity, with several features high rising frequency as the event emerges. An event is therefore represented by a set of keywords showing burst in appearance numbers [10]. The authors assume that some related words would show an increased usage as an event occurs. These techniques analyze distributions of features and detect events by grouping bursty features with identical topics. The further event detection approaches from Twitter are available for instance in the survey [11].

Twitter analysis approaches are concentrated particularly on English (sometimes also on French or on Chinese). Relatively few works are focused on the other languages. Twitter activity of the users in such languages is high and therefore common harvesting methods provided by Twitter API are sufficient to get enough of data for a further analysis. Note, that only few work about automatic event detection from Czech Twitter exists.

3   Czech Twitter Analysis and Event Detection

This section represents the main research contribution of this paper. We describe here three methods dedicated to data acquisition from Twitter with a particular focus on the proposed method based on the list of the users. Then we details the pre-procession and our event detection approach. All these tasks are necessary to build our event detection system.

3.1 Data Acquisition

The data harvesting method must fulfill the following properties:

- harvesting a "sufficient" number of tweets;
- downloading Czech Tweets;
- usage for free;
- working in real-time;
- connection to the Java programming language;
- downloading only informative messages (optional).

We describe next the different data harvesting possibilities of Twitter. We compare methods provided by Twitter API with the proposed method. We show for all the methods the maximum download speed defined by Twitter protocols. However, this speed usually differ from the real one, because the activity of the users of Twitter is not sufficient to fill these limits.

Twitter API is a publicly available library that serves to access to Twitter data. Although its simplicity, it contains numerous practical functions for user authentication, accounts management, tweet control and so on. The main functionality is provided for free and is available for every registered user. The communication with the API uses HTTP queries and the response containing the data is in JSON[5] format.

Search API Method. Search API belongs to the Twitter REST API which provides program access for reading/writing Twitter data. The functions from this library include for instance the creation of a new tweet, reading of user profile, searching a tweet or user, etc.

Search API allows queries against the indices of recent or popular tweets and behaves similarly to, but not exactly like the search feature available in web clients. This API searches against a sampling of "recent" tweets published in the past 7 days and the maximum download speed is 72,000 tweets/hour. It is possible to restrict the query by several constraints as for instance by a geolocation or by a target language.

This API is focused on relevance and not on completeness. This means that some tweets and users may be missing from the search results. The first approach, which is further evaluated and compared, uses this API. It is hereafter called as *Search API* method.

Filtered Streaming API Method. Streaming API is used for on-line Twitter monitoring (or processing). It is stated that this API represents the fastest continuous access to tweets.

There are three different streams with three different connection types available. The first one, *User stream*, is dedicated to use for providing of the data of the connected user. The second one is *Site stream*. This is an adapted version of

---

[5] JavaScript Object Notation is a lightweight format for data exchange.

the previous one for more specified users. The third one, *Public stream*, allows to get public data from different users and about different topics. We need to harvest the data from a open set of users, therefore only *Public stream* can be suitable for our task.

Twitter proposes *Sample*, *Filter* and *Firehose* connection types. The *Sample* connection provides only a sample from all the data without a good possibility to manage queries. *Firehose* connection returns all possible data and could be suitable for our task. However, this option is not free of charge. The last possibility, *Filter* connection, facilitates tweets filtering according to several critters and is free of charge. Therefore, based on the characteristics mentioned above, we can use only *Filter* connection.

The query can be, as in the case of the *Search API*, restricted by several constraints (e.g. geolocation or target language). The maximum download speed of this method is unfortunately not given. The second evaluated approach uses this API and is hereafter referred as *Filtered Streaming API (FSA)* method.

UserList Method. UserList is a Twitter functionality to allow each user to create 20 lists with the possibility to save up to 5,000 users into one list. These lists can be used to show all tweets that these users have posted and this procedure can be used with Twitter API to get all published data from 100,000 particular users.

The proposed method uses these list for harvesting of the significant amount of tweets in a given language (in Czech in our case, however the method is sufficiently general to handle the other ones). The downloaded messages should contain valuable information for data-mining and further analysis as for instance potentials events.

Our issue is now to select the representative users in order to detect appropriate tweets. Our system is designed for general event detection. Therefore it must cover the all Twitter topics by active authors from all fields. We use a small sample of interesting people provided by Czech News Agency and this sample is automatically extended by our algorithm.

The proposal of this method is based on the following observations:

– Our preliminary studies have shown that the methods provided by Twitter API are not suitable for our goal;
– about 20% of Twitter users post informative tweets, whereas the remaining 80% not [12];
– we have already a representative group of the users (sample provided by ČTK);
– this set covers a significant part of our domain of interest;
– their followers would be the users with similar interests.

Therefore, we get using the Twitter API detailed information about all the followers of our initial group. Then, we filter out all foreign (no Czech) users and we continue with the first step. When a requested number of the users is explored, the algorithm is stopped.

For every user u, it is then computed a rank R_u which is based on its number of followers Fn and the number of submitted tweets Tn as follows:

$$R_u = w.Fn + (1 - w).Tn \qquad (1)$$

where w is the importance of both criteria and was set experimentally to 0.5.

Our list is sorted by this rank and the "best" 100,000 users are added to our twitter lists for a further processing. We use Twitter API to save the data from this representative list of the 100,000 users to our data storage.

Twitter social medium is dynamic and it is modified very quickly. Therefore, this list must be periodically updated to keep actual information. It is also worth of noting that this method is language independent. This algorithm is hereafter called *UserList* method.

3.2   Pre-processing

Spam Filtering. We realize spam filtering in order to remove non informative useless tweets. These tweets are filtered with a manually defined list of entire tweets or set of rules. Table 1 shows some examples of whole tweets. The rules are based on the predefined patterns.

Table 1. Examples of the tweets to filter.

Tweet	English translation
Automatically created messages	
Přidal jsem novou zprávu na Facebook	I have added a new message on Facebook
Líbí se mi video @YouTube	I like @YouTube movie
Označil(-a) jsem video @YouTube	I have marked @YouTube movie
(Everyday) useless tweets created by the users	
Dobré odpoledne!	Good afternoon!
Jdu večeřet, dobrou chuť	I'm going to have dinner, enjoy your meal

Note that this method cannot guarantee to filter all useless tweets due to its simplicity. However, we assume that they will not be detected as events by our detection algorithm due to their not significant amount. Therefore, it is not necessary to implement more sophisticated filtering approach for the current system.

Lemmatization. Lemmatization replaces the particular (inflected) word form by its base form lemma (base form). It thus decreases the number of parameters of the system and is successfully used in many natural language processing and related tasks. We assume that lemmatization can improve the detection performance of our method. It can be useful particularly in clustering to group together appropriate words.

Following the definition from the Prague Dependency Treebank (PDT) 2.0 [13] project, we use only the first part of the lemma. This is a unique identifier of the lexical item (e.g. infinitive for a verb), possibly followed by a digit to disambiguate different lemmas with the same base forms. For instance, the Czech word "třeba", having the identical lemma, can signify *necessary* or *for example* depending on the context. This is in the PDT notation differentiated by two lemmas: "třeba-1" and "treba-2". The second part containing additional information about the lemma, such as semantic or derivational information, is not taken into account in this work.

Note that we also envisaged to use stemming, an analogical process to lemmatization, which also reduces inflected words to their root form as called "stem". The stem need not be identical to the morphological root of the word. It is usually sufficient that related words map to the same stem, even if this stem is not in itself a valid root. However, based on our previous experiments [14], we decided to use lemmatization, rather than stemming.

Non-significant Word Filtering. Non-significant (or stop) words are words with high frequencies which have in a sentence rather grammatical meaning as for instance prepositions or conjunctions. Our filtering is based on a manually defined list. We will implement more sophisticated method based for instance on Part-of-Speech (POS) tags in the further version. However, we assume that this improving will play only marginal role for event detection.

3.3 Event Detection

One-Pass Clustering. We propose a one-pass clustering method to extract the events. We consider that we have in real-time the filtered and lemmatized tweets which can represent (due to the UserList method) probably the events. We transform every tweet into a binary representation using a bag of words method, which represents its unique location in n-dimensional space. Then the clustering algorithm is as follows:

1. take an (unprocessed) tweet;
2. calculate the cosine distance between a vector representing this tweet and all the others;
3. choose a closest tweet (or cluster of tweets if any) and group them together (the maximum allowed distance is given by the threshold *Th*;
4. repeat the two previous operations (*go to step 1*) till all tweets are processed.

The clusters created by this algorithm represent the events. Of course, the clustering does not guarantee that the created clusters represent only the events. This should be done by the pre-processing:

– UserList data acquisition method harvests particularly informative tweets which contains mainly the events;
– spam filtering step removes several non informative useless tweets (no events).

We also define a parameter T which indicates a time period for the clustering. We assume that different events will be produced at different "speed" (different activities of Twitter users). For instance, information about the winner of the football championship can be quicker (more contributions in a short period) than information about a new director of some company.

It is worth of noting, that we have also considered a *gradient* of the word frequencies in some event clusters. It means, that a significant increase of the presence of particular word/words indicates probably an event. However, this approach did not work because of the small activity of the users on Czech Twitter.

Results Representation. The results of the clustering are thus the groups of tweets with some common words. This group is represented by the *most signif- icant* tweet. This tweet is defined as a message with the maximum of common words and the minimum of the other words. This representation is used due to the effort to use an answer in natural language, instead of a list of key-words or a phrase.

4 System Description

We describe in the following text the main properties and general architecture of the proposed event detection system. The system is implemented under Java 8 programming language and has a modular architecture. It is composed of three main functional units (*Tweet Stream Analysis*, *Preprocessing* and *Event Detection*) which are further decomposed into six modules, as shown in Fig. 1.

M1: Data Acquisition. The first module, *Data Acquisition*, is used for on- line storage of appropriate information from Twitter stream in Czech language with high download speed. We integrate the proposed UserList method into this module to harvest a sufficient number of tweets. The output of this module are raw tweets without any post-processing.

M2: Spam Filtering. The second one is *Spam filtering* module. It removes tweets with useless information (so called "spam") using rule-based approach. The input are raw tweets collected by the previous module *M1* and the output consists of a set of partially filtered tweets.

M3: Lemmatization. The third, *Lemmatization* module is used for word nor- malization to decrease the number of feature in the system. The input are par- tially filtered tweets provided by the module *M2* and this module outputs lem- matized text.

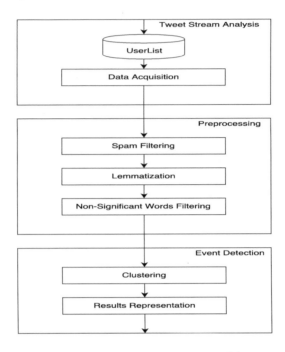

Fig. 1. Architecture of the system [1].

M4: Non-significant Word Filtering. The next one is *Non-significant word filtering* module. While the previous filtering was at the tweet level, this one filters the words and is thus used to remove non-significant words which could decrease the detection performance.

M5: Clustering Module. The *Clustering* module is used to discover the events. We group together the tweets with similar content using a one-pass clustering method. The final decision about detected events is based on thresholding. The input of this module is the text of lemmatized and filtered tweets and this module provides information about the discovered events.

M6: Results Representation. This last module is used to show the detected events to the users in an acceptable form. One tweet representing the detected event is presented to the user in this step.

5   Evaluation

We describe next the experiments realized to evaluate the proposed Twitter harvesting method. This section further details the results of our event detection approach.

5.1 Data Acquisition

Comparison of the Czech and French Twitter Activity. In this experiment, we confirm our claim that the activity of the Czech Twitter is significantly lower than in the case of the other languages. We have chosen French Twitter and the *Search API* method (see Sect. 3.1) for such comparison.

First, we have shown that it is not possible to use language constraints to obtain only the Czech tweets. However, the Czech constraint is missing and there is available only "sk" field which contains Czech and Slovak tweets together.

Therefore, we have decided to filter tweets according to the geolocation. We have chosen a square region, covering most of the territories of the Czech Republic and France, as our area of interest. We have analyzed the download rate in interval from 22 to 29 August 2015. Figure 2 shows the results of this analysis.

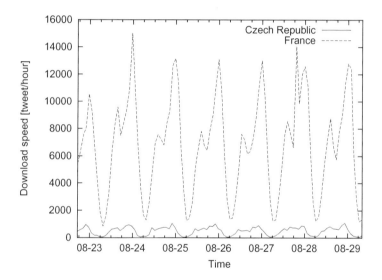

Fig. 2. Comparison of Czech and French Twitter activity [1].

This figure shows that the activity of French Twitter is more than 10 × higher than the Czech Twitter. The average of the Czech download rate is about 495 tweets/hour. However, after a detailed examination, we have identified that only less than 20% of tweets are written in Czech languages.

Unfortunately, this number is insufficient for a successful further analysis as for instance for event detection in real-time. Therefore, we must analyze the other approaches for data acquisition.

Comparison of the Different Data Acquisition Methods. In this experiment, we compare the download speed of two standard methods provided by the Twitter API (namely *Search API* and *Filtered Streaming API (FSA)* methods with the proposed *UserList* approach (see Sect. 3.1). We have thus executed all these methods in the same two day period and then we have calculated the average value for one hour.

Table 2. Comparison of the download speed of the different methods on the Czech Twitter.

Method	Search API	FSA	UserList *(proposed)*
Download speed [Tweets/hour]	43.5	56.6	**330.3**

The results of this experiment are shown in Table 2. This table shows that the proposed method provides about 6 times more data than the standard methods provided by Twitter API. Based on these results we have chosen the *UserList* approach to integrate into our event detection system.

5.2   Event Detection

Event Corpus. We created a small corpus of Czech tweets to evaluate the event detection method. It is composed of 536 Tweets collected during one hour using the proposed *UserList* data acquisition method described previously. This corpus contains 4 978 word tokens being 3 532 different words and 39 events. It was annotated by two different annotators with annotator agreement 87.9%. The ambiguous cases was decided by the third experienced annotator.

Detection Results. Figure 3 shows the results of the proposed event detection method depending on the different acceptation threshold *Th*. The standard recall, precision and f-measure (*F1*) metrics [15] are used for evaluation of this experiment. This figure shows that the best detection results are obtained with the threshold value *Th = 0.6*. This figure further shows that precision values are significantly higher than the values of recall.

The obtained results are promising, although the resulting f-measure is not much high and should be further improved. It could be done by the extension of the corpus and by adaptation of the clustering method. On the larger corpus, it will be possible to use longer time periods (two, four, six hours, etc.) to detect the events that are written at different speeds, which cannot be detected on the current version of the corpus. We also envisage to use word embeddings for better representation of words in the clustering method instead of the simple bag of words. This representation is much more precious and it should create better clusters of the tweets.

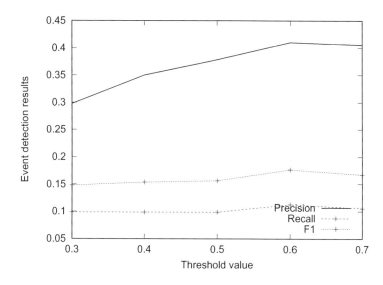

Fig. 3. Event detection results dependent on different acceptance threshold Th.

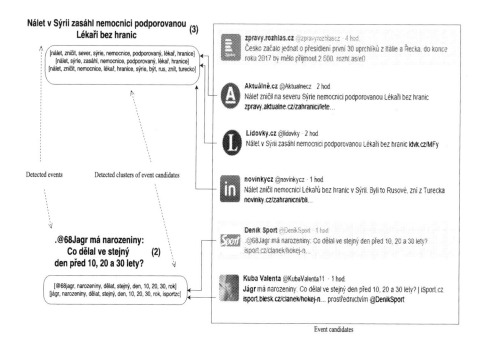

Fig. 4. Event detection example (time period $T = 2h$ and acceptance threshold $Th = 0.5$). The rectangle on the right contains six tweets that were saved by our acquisition method. The left "bubbles" show the results of our clustering (two groups containing three and two tweets). The representative tweets are chosen (marked by the bold text on the left side) to be presented to the user [1].

Detection Example. The sample results are depicted in Fig. 4. This figure shows that six tweets are saved by our acquisition method (right rectangle). They are then clustered into two groups containing three and two tweets (left "bubbles"). Finally, one representative tweet is chosen from both clusters to be presented to the user (bold text left).

6   Contributions

The main contributions of this paper are summarized below:

- proposing an algorithm for real-time tweet acquisition adapted to the Czech Twitter based on the *UserLists*;
- proposing an algorithm for event detection from Czech Twitter;
- preliminary evaluating of the proposed methods on a small Czech Twitter corpus of the events.

7   Conclusions and Future Work

The main goal of this paper was to propose an approach to harvest messages from Twitter in Czech language with high download speed and a method to detect automatically important events. The proposed harvesting method uses user lists to discover potentially interesting tweets to harvest. We have experimentally shown that this method is efficient because it harvests about 6 times more data than the two other approaches provided by the Twitter API. We further created a small Czech Twitter corpus annotated with the events. We evaluated our event detection method on this corpus. We experimentally shown that the results of the event detection are promising.

Our first perspective consists in the extension of the corpus. On the larger corpus, it will be possible to use longer time periods (two, four, six hours, etc.) to detect the events that are written at different speeds, which cannot be detected on the current version. We also plan to use word embeddings for better representation of words in the clustering method instead of the simple bag of words. This representation is much more precious and it should create better clusters of the tweets. We would like also improve clustering method using more sophisticated semantic similarity functions. The whole system is language independent. Therefore, the last perspective consists in evaluation of the system on other (particularly European) languages.

Acknowledgements. This work has been partly supported by the project LO1506 of the Czech Ministry of Education, Youth and Sports and by Grant No. SGS-2016-018 Data and Software Engineering for Advanced Applications.

References

1. Král, P., Rajtmajer, V.: Real-time data harvesting method for Czech twitter. In: in 9th International Conference on Agents and Artificial Intelligence (ICAART 2017), Porto, Portugal, pp. 259–265. SciTePress (2017)
2. Pak, A., Paroubek, P.: Twitter as a corpus for sentiment analysis and opinion mining. In: LREc, vol. 10, pp. 1320–1326 (2010)
3. Kouloumpis, E., Wilson, T., Moore, J.D.: Twitter sentiment analysis: the good the bad and the OMG! In: ICWSM, vol. 11, pp. 538–541 (2011)
4. Yardi, S., Boyd, D.: Dynamic debates: an analysis of group polarization over time on Twitter. Bull. Sci. Technol. Soc. **30**, 316–327 (2010)
5. Java, A., Song, X., Finin, T., Tseng, B.: Why we Twitter: an analysis of a microblogging community. In: Zhang, H., Spiliopoulou, M., Mobasher, B., Giles, C.L., McCallum, A., Nasraoui, O., Srivastava, J., Yen, J. (eds.) SNAKDD/WebKDD 2007. LNCS (LNAI), vol. 5439, pp. 118–138. Springer, Heidelberg (2009). https://doi.org/10.1007/978-3-642-00528-2_7
6. Sakaki, T., Okazaki, M., Matsuo, Y.: Earthquake shakes Twitter users: real-time event detection by social sensors. In: Proceedings of the 19th International Conference on World Wide Web, pp. 851–860. ACM (2010)
7. Earle, P.S., Bowden, D.C., Guy, M.: Twitter earthquake detection: earthquake monitoring in a social world. Ann. Geophys. **54**, 708–715 (2012)
8. Li, C., Sun, A., Datta, A.: Twevent: segment-based event detection from tweets. In: Proceedings of the 21st ACM International Conference on Information and Knowledge Management, pp. 155–164. ACM (2012)
9. Huang, S., Yang, Y., Li, H., Sun, G.: Topic detection from microblog based on text clustering and topic model analysis. In: 2014 Asia-Pacific Services Computing Conference (APSCC), pp. 88–92. IEEE (2014)
10. Kleinberg, J.: Bursty and hierarchical structure in streams. In: Proceedings of the Eighth ACM SIGKDD International Conference on Knowledge Discovery and Data Mining, pp. 91–101. ACM (2002)
11. Atefeh, F., Khreich, W.: A survey of techniques for event detection in Twitter. Comput. Intell. **31**, 132–164 (2015)
12. Naaman, M., Boase, J., Lai, C.H.: Is it really about me?: message content in social awareness streams. In: Proceedings of the 2010 ACM Conference on Computer Supported Cooperative Work, pp. 189–192. ACM (2010)
13. Zeman, D., Dušek, O., Mareček, D., Popel, M., Ramasamy, L., Štěpánek, J., Žabokrtský, Z., Hajič, J.: Hamledt: harmonized multi-language dependency treebank. Lang. Res. Eval. **48**, 601–637 (2014)
14. Hrala, M., Král, P.: Evaluation of the document classification approaches. In: Burduk, R., Jackowski, K., Kurzynski, M., Wozniak, M., Zolnierek, A. (eds.) CORES 2013. AISC, vol. 226, pp. 877–885. Springer, Heidelberg (2013). https://doi.org/10.1007/978-3-319-00969-8_86
15. Powers, D.: Evaluation: from precision, recall and f-measure to ROC, informedness, markedness & correlation. J. Mach. Learn. Technol. **2**, 37–63 (2011)

Variants of Independence Detection in SAT-Based Optimal Multi-agent Path Finding

Pavel Surynek[1]($\boxtimes$), Jiří Švancara[2], Ariel Felner[3], and Eli Boyarski[4]

[1] National Institute of Advanced Industrial Science and Technology (AIST), Tokyo, Japan
pavel.surynek@aist.go.jp
[2] Faculty of Mathematics and Physics, Charles University, Prague, Czechia
jiri.svancara@mff.cuni.cz
[3] Ben Gurion University, Beer-Sheva, Israel
felner@bgu.ac.il
[4] Bar-Ilan University, Ramat-Gan, Israel
eli.boyarski@gmail.com

Abstract. In multi-agent path finding (MAPF) on graphs, the task is to find paths for distinguishable agents so that each agent reaches its unique goal vertex from the given start while collisions between agents are forbidden. A cumulative objective function is often minimized in MAPF. The main contribution of this paper consists in integrating independence detection technique (ID) into a compilation-based MAPF solver that translates MAPF instances into propositional satisfiability (SAT). The independence detection technique in search-based solvers tries to decompose a given MAPF instance into instances consisting of small groups of agents with no interaction across groups. After the decomposition phase, small instances are solved independently and the solution of the original instance is combined from individual solutions to small instances. The presented experimental evaluation indicates significant reduction of the size of instances translated to the target SAT formalism and positive impact on the overall performance of the solver.

Keywords: Multi-agent path-finding (MAPF) · Independence detection (ID)
Propositional satisfiability (SAT) · Cost optimality · Makespan optimality
Sum-of-costs optimality · SAT encodings · Path-finding on grids

1 Introduction

Multi-agent path finding (MAPF) represents a task of finding collision free paths for a set of mobile agents where each agent is assigned unique start and goal positions [13, 19, 22, 27]. An environment with agents is often abstracted by *undirected graph* in the literature [17, 36]. Agents in this abstraction are represented by items placed in vertices of the graph while edges represent passable regions. At most one agent can be placed in each vertex to model physical space occupancy by agents. Agents can traverse a single edge at each time step.

© Springer International Publishing AG, part of Springer Nature 2018
J. van den Herik et al. (Eds.): ICAART 2017, LNAI 10839, pp. 116–136, 2018.
https://doi.org/10.1007/978-3-319-93581-2_7

Various movement schemes exist for MAPF on graphs. Often an agent can move into an unoccupied neighbor not entered by another agent [28] – this will be called *move-to-unoccupied* variant. This variant requires at least one vertex in the graph unoccupied to be able to perform some movements at all. Other movement schemes include chains of agents moving simultaneously with only the leader entering the unoccupied vertex or cases with no vacant vertex in the graph where rotations along non-trivial cycles are allowed [32]. We base our presentation on the *move-to-unoccupied* variant. Let us note that techniques shown in this paper are generic across all these movement schemes.

The MAPF problem and its variants are strongly practically motivated. Applications range from navigation of multiple mobile robots [5, 8] through traffic optimization [12, 15] to movement planning in computer games [34]. We refer the reader to various studies such as [19, 20] for the detailed survey of applications.

1.1 Optimality in MAPF

We address *optimal* MAPF in which we search for paths that are optimal with respect to a given objective. The two basic cumulative objectives studied in the literature are *sum-of-costs* [19, 25] and *makespan* [29].

The **sum-of-costs objective** assumes that unit costs are assigned to move and wait actions. The cost of plan is the sum of action costs along all the paths and over all agents. The aim is to obtain a plan with the minimum cost.

Under the **makespan objective,** the aim is to obtain a plan that can be executed in as short as possible time while each movement consumes 1 unit of time. In other words, we need the longest path out of all the paths to be as short as possible.

As we will show later, there may be situations where the increase in the sum-of-costs leads to a shorter makespan and vice versa.

A feasible solution of MAPF can be found in polynomial time [13, 35]. Adding any of the discussed objectives renders the decision version of MAPF (a *yes/no question* if a given MAPF has a solution of specified makespan/sum-of-costs) to be NP-complete [16, 28, 32].

We will keep the further description around the sum-of-costs variant but it is important to note that the presented techniques apply for the makespan variant as well.

1.2 Contributions to SAT-Based MAPF

Techniques for solving MAPF optimally include translation of the decision version into *propositional formula* [10, 11]. The formula is *satisfiable* if and only if the instance of MAPF is solvable for a given value of the objective function. Assuming that satisfiability of such formula is a non-decreasing function of the value of objective function, the optimum can be obtained by querying the satisfiability multiple times. A trivial strategy of increasing the value of objective function by one turned out to be a good choice [30] in many cases (thanks to the non-uniform difficulty of each query).

Satisfiability of the formula can be decided by an off-the-shelf SAT solver [2, 6] which is one of the advantages of the SAT-based approach. All advanced techniques developed in recent SAT solvers [4] can be employed for solving MAPF in this way.

The most significant bottleneck of all existing SAT-based algorithms for MAPF is the large size and combinatorial difficulty of the target propositional formula that grow significantly with the increasing number of agents as well as with growing size of the underlying graph. This kind of growth of combinatorial difficulty has already been addressed by Standley [24] in his search-based optimal MAPF solving algorithm. Standley described various variants of a method called *independence detection* that tries to determine the smallest possible groups of agents for which paths can be found independently of other groups.

Our contribution consists in integrating two variants of independence detection – *simple independence detection* (SID) and *independence detection* (ID) – with MDD-SAT – the most recent SAT-based MAPF solver [30]. As there are differences in how the original Standley's search-based algorithm and SAT-based approach work, we suggested modifications to ID to be compatible with the SAT-based approach. Our new solvers are called MDD-SAT+SID and MDD-SAT+ID following the notation of [24]. Conducted experiments demonstrate similar performance benefit as in the case of original application of SID and ID in the search-based approach.

This paper is an extension of [31]. We describe in more detail the encoding of MAPF to a Boolean formula. In addition to [31] we also present experimental evaluation of MDD-SAT+SID. The paper is organized as follows. After the formal introduction of the MAPF problem a brief exposition of related work is done. Then, the Boolean encoding of MAPF is presented, also the original SID and ID are recalled and their integration with the SAT-based approach is presented. Finally, an experimental evaluation with grids and large maps is presented.

2 MAPF Definition

An arbitrary **undirected graph** can be used to model the environment where agents are moving. Let $G = (V, E)$ be such a graph where $V = \{v_1, v_2, \ldots, v_n\}$ is a finite set of vertices and $E \subseteq \binom{V}{2}$ is a set of edges.

The placement of agents in the environment is modeled by assigning them vertices of the graph. Let $A = \{a_1, a_2, \ldots, a_m\}$ be a finite set of *agents*. Then, an arrangement of agents in vertices of graph G will be fully described by a *location* function $\alpha: A \rightarrow V$; the interpretation is that an agent $a \in A$ is located in a vertex $\alpha(a)$. At most **one agent** can be located in each vertex; that is α is uniquely invertible.

Definition 1 (MAPF). An instance of *multi-agent path-finding* problem is a quadruple $\Sigma = [G = (V, E), A, \alpha_0, \alpha_+]$ where location functions α_0 and α_+ define the initial and the goal arrangement of a set of agents A in G respectively.   □

The dynamicity of the model assumes a discrete time divided into time steps. An arrangement α_i at the i-th time step can be transformed by a transition action which instantaneously moves agents in the non-colliding way to form a new arrangement α_{i+1}. The transition between α_i and α_{i+1} must satisfy the following *validity conditions*:

(1) $\forall a \in A$ either $\alpha_i(a) = \alpha_{i+1}(a)$ or $\{\alpha_i(a), \alpha_{i+1}(a)\} \in E$ holds
     (agents move along edges or wait at their current location),
(2) $\forall a \in A\, \alpha_i(a) \neq \alpha_{i+1}(a) \Rightarrow \alpha_i^{-1}(\alpha_{i+1}(a)) = \bot$
     (agents move to vacant vertices only), and
(3) $\forall a, b \in A\, a \neq b \Rightarrow \alpha_{i+1}(a) \neq \alpha_{i+1}(b)$
     (no two agents enter the same target/unique invertibility of resulting arrangement).

The task in MAPF is to transform α_0 using above valid transitions to α_+. An illustration of MAPF and its solution is depicted in Fig. 1.

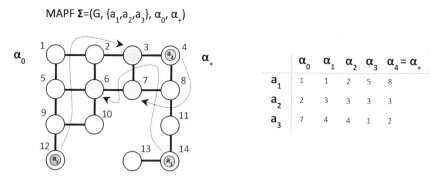

Fig. 1. An example of a MAPF instance from [31] with three agents a_1, a_2, and a_3 (left). A solution of the instance is shown (right).

Definition 2 (MAPF solution). A *solution* for MAPF instance $\Sigma = [G, A, \alpha_0, \alpha_+]$ is a sequence of arrangements $[\alpha_0, \alpha_1, \alpha_2, \ldots, \alpha_\mu]$ where $\alpha_\mu = \alpha_+$ and α_{i+1} is a result of valid transition from α_i for every $= 1, 2, \ldots, \mu - 1$.   □

Makespan μ is the total number of time steps until the last agent reaches its destination. *Sum-of-costs* denoted ξ is the sum of path costs per individual agents. Each action (including wait) of an agent before it reaches its goal has unit cost.

2.1 Makespan vs. Sum-of-Costs

There exists an instance in which all the sum-of-costs optimal solutions are not makespan optimal. Similarly, none of the makespan optimal solution is sum-of-costs optimal there (see Fig. 2 for illustration).

In the SAT-based optimal MAPF solver described below, a proper relation between makespan and sum-of-costs need to be found as both objectives are bounded during search. We need to ensure that smallest cost found under the given makespan bound is optimal (see [30] for more detailed discussion).

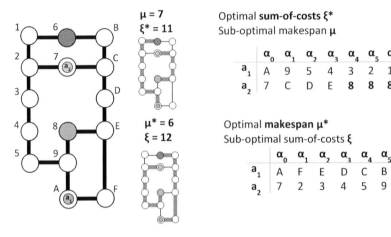

Fig. 2. An instance of the MAPF problem from [31] in which no *makespan* optimal solution is sum-of-costs optimal and no *sum-of-costs* optimal solution is makespan optimal.

3 Related Work

Many other successful algorithms exist for the optimal MAPF solving. The state-of-the-art search-based algorithms (though there is no universal winner) include *increasing cost tree search* - ICTS [19], *conflict base search* - CBS [20], and *improved CBS* – ICBS [7]. These algorithms excel in setups with relatively few agents on large maps.

Another research direction is represented by methods based on reduction of the MAPF problem to another formalism. Except the SAT as a target formalism, successful attempts to reduce MAPF to *constraint optimization problem* [18], *inductive logic programming* [33], and *answer set programming* [9] have been made. These approaches (the SAT approach including) can be generally characterized by a high performance in MAPFs with small underlying graph densely populated with agents. This is a natural outcome of the maturity of solvers used to solve hard combinatorial problems in the target formalism.

Recently new research directions driven by applications have been identified in the MAPF context. For example, it is not always necessary to distinguish between individual agents – see [14] for detailed survey.

4 SAT Encoding for Optimal Sum-of-Costs

In this paper, we follow the algorithm solving sum-of-cost optimal MAPF via reduction to SAT presented in [30].

The basic approach in solving MAPF via SAT is to create a *time expansion graph* (denoted TEG) [29]. A TEG is a directed acyclic graph (DAG). First, the set of vertices of the underlying graph G are duplicated for all time-steps from 0 up to the given bound μ. Then, possible actions (move along edges or wait) are represented as directed edges between successive time steps. Formally a TEG is defined as follows:

Definition 3 (TEG). *Time expansion graph of depth* μ *for underlying graph* (V, E) *is a* digraph (V_μ, E_μ) where $V_\mu = \left\{ u_j^t | t = 0, 1, \ldots, \mu \wedge u_j \in V \right\}$ and $E_\mu = \{ (u_j^u, u_k^{t+1}) |$ $t = 0, 1, \ldots, \mu - 1 \wedge (\{u_j, u_k\} \in E \vee j = k) \}$. $\qquad\qquad\square$

The encoding for MAPF introduces propositional variables and constraints for a single time-step t in order to represent any possible arrangement of agents at time t. Given a desired makespan μ, the formula represents the question of whether there is a solution in the TEG of μ time steps. The search for optimal makespan is done by iteratively incrementing $\mu (= 0, 1, 2 \ldots)$ until a satisfiable formula is obtained.

To find the optimal sum-of-costs solution, we use similar technique as with optimal makespan solution. The sequence of decision problems is whether there exists a solution of a given sum-of-cost ξ. However, encoding this decision problem is more challenging than the makespan case, because one needs to both bound the sum-of-costs, but also to predict how many time expansions are needed. We address this challenge by using two key techniques described next: (1) Cardinality constraint for bounding ξ and (2) Bounding the Makespan.

4.1 Cardinality Constraint for Bounding ξ

The SAT literature offers a technique for encoding a *cardinality constraint* [3, 21], which allows calculating and bounding a numeric cost within the formula. Formally, for a bound $\lambda \in \mathbb{N}$ and a set of propositional variables $X = \{x_1, x_2, \ldots, x_k\}$ the cardinality constraint $\leq_\lambda \{x_1, x_2, \ldots, x_k\}$ is satisfied iff the number of variables from the set X that are set to TRUE is $\leq \lambda$.

In our SAT encoding, we bound the sum-of-costs by mapping every agent's action to a propositional variable, and then encoding a cardinality constraint on these variables. Thus, one can use the general structure of the makespan SAT encoding (which iterates over possible makespans), and add such a cardinality constraint on top.

4.2 Bounding the Makespan for the Sum of Costs

We compute how many time expansions (μ) are needed to guarantee that if a solution with sum-of-costs ξ exists then it will be found. In other words, in our encoding, the values we give to ξ and μ must fulfill the following requirement:

R1: *all possible solutions with sum-of-costs* ξ *must be possible for a makespan of at most* μ.

To find a μ value that meets R1, we require the following definitions. Let $\xi_0(a_i)$ be the cost of the shortest individual path for agent a_i, and let $\xi_0 = \sum_{a_i \in A} \xi_0(a_i)$. ξ_0 was called the sum of individual costs (SIC) [19]. ξ_0 is an admissible heuristic for optimal sum-of-costs search algorithms, since ξ_0 is a lower bound on the minimal sum-of-costs. ξ_0 is calculated by relaxing the problem by omitting the other agents. Similarly, we define $\mu_0 = \max_{a_i \in A} \xi_0(a_i)$. μ_0 is length of the *longest* of the shortest individual paths and is thus a lower bound on the minimal makespan. Finally, let Δ be the extra cost over SIC (as done in [19]). That is, let $\Delta = \xi - \xi_0$.

Proposition 1. *For makespan μ of any solution with sum-of-costs ξ, R1 holds for $\mu \leq \mu_0 + \Delta$.*

Proof Outline: The worst-case scenario, in terms of makespan, is that all the Δ extra moves belong to a single agent. Given this scenario, in the worst case, Δ is assigned to the agent with the largest shortest path. Thus, the resulting path of that agent would be $\mu_0 + \Delta$, as required. □

Using Proposition 1, we can safely encode the decision problem of whether there is a solution with sum-of-costs ξ by using $\mu = \mu_0 + \Delta$ time expansions, knowing that if a solution of cost ξ exists then it will be found within $\mu = \mu_0 + \Delta$ time expansions. In other words, Proposition 1 shows relation of both parameters μ and ξ which will be both changed by changing Δ. Algorithm 1 summarizes our optimal sum-of-costs algorithm. In every iteration, μ is set to $\mu_0 + \Delta$ and the relevant TEGs (described below) for the various agents are built. Next a decision problem asking whether there is a solution with sum-of-costs ξ and makespan μ is queried. The first iteration starts with $\Delta = 0$. If such solution exists, it is returned. Otherwise ξ is incremented by one, Δ and consequently μ are modified accordingly and another iteration of SAT consulting is activated.

Algorithm 1. SAT consult illustrating the increase in Δ.

```
MAPF-SAT  (MAPF Σ = (G = (V, E), A, α₀, α₊))
   μ₀ = max_{aₜ∈A} ξ₀(aᵢ),  Δ = 0
   while Solution not found do
      μ = μ₀ + Δ
      for each agent aᵢ do
         Build TEG_i(μ)
      end
      Solution = Consult-SAT-Solver(Σ, μ, Δ)
      if Solution not found then
         Δ++
      end
   end
   return (Solution)
end
```

This algorithm clearly terminates for solvable MAPF instances as we start seeking a solution of $\xi = \xi_0 (\Delta = 0)$ and increment Δ (which increments ξ and μ as well) to all possible values. The unsolvability of an MAPF instance can be checked separately by a polynomial-time complete sub-optimal algorithm such as PUSH-AND-ROTATE [35].

4.3   Efficient Use of the Cardinality Constraint

The complexity of encoding a cardinality constraint depends linearly in the number of constrained variables [21, 23]. Since each agent a_i must move at least $\xi_0(a_i)$, we can reduce the number of variables counted by the cardinality constraint by only counting the variables corresponding to extra movements over the first $\xi_0(a_i)$ movement a_i makes. We implement this by introducing a TEG for a given agent a_i (labeled $TEGi$).

TEG_i differs from TEG (Definition 3) in that it distinguishes between two types of edges: E_i and $F_i \cdot E_i$ are (directed) edges whose destination is at time step $\leq \xi_0(a_i)$. These are called standard edges. F_i denoted as extra edges are directed edges whose destination is at time step $\geq \xi_0(a_i)$. Figure 3 shows an underlying graph for agent a_1 (left) and the corresponding TEG_1. Note that the optimal solution of cost 2 is denoted by the diagonal path of the TEG. Edges that belong to F_i are those that their destination is time step 3 (dotted lines). The key in this definition is that the cardinality constraint would only be applied to the extra edges, that is, we will only bound the number of extra edges (they sum up to Δ) making it more efficient. There are various possibilities to define what happens to an agent when it reaches the goal (disappears, waits etc.). In all cases, edges in TEGs corresponding to wait actions at the goal are not marked as extra. Importantly, our SAT approach is robust across all these variants.

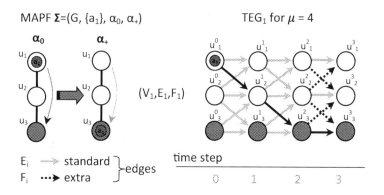

Fig. 3. A TEG for an agent that needs to go from u_1 to u_3.

4.4   Detailed Description of the SAT Encoding

Agent a_i must go from its initial position to its goal within TEG_i. This simulates its location in time in the underlying graph G. That is, the task is to find a path from $a_0^0(a_i)$ to $a_+^\mu(a_i)$ in TEG_i. The search for such a path will be encoded within the Boolean formula. Additional constraints will be added to capture all movement constraints such as *collision avoidance* etc. And, of course, we will encode the cardinality constraint that the number of extra edges must be exactly Δ.

We want to ask whether a sum-of-costs solution of ξ exists. For this we build TEG_i for each agent $a_i \in A$ of depth $\mu_0 + \Delta$. We use V_i to denote the set of vertices in TEG_i

that agent a_i might occupy during the time steps. Next we introduce the Boolean encoding (denoted BASIC-SAT) which has the following Boolean variables:

1. $\chi_j^t(a_i)$ for every $t \in \{0, 1, \ldots, \mu\}$ and $u_j^t \in V_i$ – Boolean variable of whether agent a_i is in vertex v_j at time step t.

2. $\mathcal{E}_{j,k}^t(a_i)$ for every $t \in \{0, 1, \ldots, \mu - 1\}$ and $\left(u_j^t, u_k^{t+1}\right) \in (E_i \cup F_i)$ – Boolean variable that model transition of agent a_i from vertex v_j to v_k through any edge (standard or extra) between time steps t and $t + 1$ respectively.

3. $C^t(a_i)$ for every $t \in \{0, 1, \ldots, \mu - 1\}$ such that there exist $u_j^t \in V_i$ and $u_k^{t+1} \in V_i$ with $\left(u_j^t, u_k^{t+1}\right) \in F_i$ – Boolean variables that model cost of movements along extra edges (from F_i) between time steps t and $t + 1$.

We now introduce constraints on these variables to restrict illegal values as defined by our variant of MAPF. Other variants may use a slightly different encoding but the principle is the same. Let $T_\mu = \{0, 1, \ldots, \mu - 1\}$. Several groups of constraints are introduced for each agent $a_i \in A$ as follows:

C1: If an agent appears in a vertex at a given time step, then it must follow through exactly one adjacent edge into the next time step. This is encoded by the following two constraints, which are posted for every $t \in T_\mu$ and $u_j^t \in V_i$.

$$\chi_j^t(a_i) \Rightarrow \bigvee\nolimits_{\left(u_j^t, u_k^{t+1}\right) \in (E_i \cup F_i)} \mathcal{E}_{j,k}^t(a_i) \tag{1}$$

$$\bigwedge\nolimits_{\left(u_j^t, u_k^{t+1}\right), \left(u_j^t, u_l^{t+1}\right) \in (E_i \cup F_i) \wedge k < l} \neg \mathcal{E}_{j,k}^t(a_i) \vee \neg \mathcal{E}_{j,l}^t(a_i) \tag{2}$$

C2: Whenever an agent occupies an edge it must also enter it before and leave it at the next time-step. This is ensured by the following constraint introduced for every $t \in T_\mu$ and $\left(u_j^t, u_k^{t+1}\right) \in (E_i \cup F_i)$.

$$\mathcal{E}_{j,k}^t(a_i) \Rightarrow \chi_j^t(a_i) \wedge \chi_k^{t+1}(a_i) \tag{3}$$

C3: The target vertex of any movement except wait action must be empty. This is ensured by the following constraint introduced for every $t \in T_\mu$ and $\left(u_j^t, u_k^{t+1}\right) \in (E_i \cup F_i)$ such that $j \neq k$.

$$\mathcal{E}_{j,k}^t(a_i) \Rightarrow \bigwedge\nolimits_{a_l \in A \wedge a_l \neq a_i \wedge u_k^{t+1} \in V_l} \chi_k^{t+1}(a_l) \tag{4}$$

C4: No two agents can appear in the same vertex at the same time step (although the previous constraint ensures that an agent does not collide with an agent currently residing in a vertex it does not prevent simultaneous entering of the same vertex by multiple agents). That is the following constraint is added for every $t \in T_\mu$ and pair of agents $a_i, a_l \in A$ such that $i \neq l$.

$$\bigwedge_{u_j^t \in V_i \cap V_l} \neg \chi_j^t(a_i) \vee \neg \chi_j^t(a_l) \tag{5}$$

C5: Whenever an extra edge is traversed the cost needs to be accumulated. In fact, this is the only cost that we accumulate as discussed above. This is done by the following constraint for every $t \in T_\mu$ and extra edge $\left(u_j^t, u_k^{t+1} \right) \in F_i$.

$$\mathcal{E}_{j,k}^t(a_i) \Rightarrow C^t(a_i) \tag{6}$$

C6: **Cardinality Constraint.** Finally, the bound on the total cost needs to be introduced. Reaching the sum-of-costs of ξ corresponds to traversing exactly Δ extra edges from F_i. The following cardinality constrains ensures this:

$$\leq_\Delta \left\{ C^t(a_i) | i = 1, 2, \ldots, n \wedge t = 0, 1, \ldots \mu - 1 \wedge \left\{ \left(u_j^t, u_k^{t+1} \right) \in F_i \right\} \neq \emptyset \right\} \tag{7}$$

The resulting Boolean formula that is a conjunction of C1 ... C7 will be denoted as $\mathcal{F}_{BASIC}(\Sigma, \mu, \Delta)$ and is the one that is consulted by Algorithm 1.

The following proposition summarizes the correctness of our encoding.

Proposition 2. *MAPF* $\Sigma = (G = (V, E), A, \alpha_0, \alpha_+)$ *has a sum-of-costs solution of* ξ *if and only if* $\mathcal{F}_{BASIC}(\Sigma, \mu, \Delta)$ *is satisfiable. Moreover, a solution of MAPF* Σ *with the sum-of-costs of* ξ *can be extracted from the satisfying valuation of* $\mathcal{F}_{BASIC}(\Sigma, \mu, \Delta)$ *by reading its* $\chi_j^t(a_i)$ *variables.*

Proof: The direct consequence of the above definitions is that a valid solution of a given MAPF Σ corresponds to non-conflicting paths in the TEGs of the individual agents. These non-conflicting paths further correspond to satisfying the variable assignment of $\mathcal{F}_{BASIC}(\Sigma, \mu, \Delta)$, i.e., that there are Δ extra edges in TEGs of depth $\mu = \mu_0 + \Delta$. $\qquad\square$

As discussed in [30], the limitation of BASIC-SAT encoding is its size which is implied by the size of the time expanded graph. To mitigate this limitation Surynek et al. took inspiration from another successful search-based solver called *increasing cost tree search* (ICTS) [19]. Vertices whose sum of distances from $a_0^0(a_i)$ and $a_+^\mu(a_i)$ in TEG$_i$ is greater than μ can never be visited by a_i in any optimal solution or else a_i would not have enough time steps to reach $a_+^\mu(a_i)$. Omitting those vertices from TEGs that are too far in the aforementioned sense would not compromise soundness of the solving process but would lead to a smaller formula. In [30], this version of TEGs where unreachable vertices are omitted is called MDD and corresponding formula is denoted as $\mathcal{F}_{MDD}(\Sigma, \mu, \Delta)$. When referring to MDD-SAT solver we assume the version with MDDs.

Using MDDs can rule out many vertices that would be normally considered in standard time expansions. Experiments confirmed that MDDs enabled using the SAT-based approach even for large MAPF instances for which the size of encodings without MDD was prohibitive.

5   Independence Detection

Our major aim is to increase performance of the SAT-based MAPF solver by reducing the number of agents needs to be considered at once. This has been successfully done in search based methods via a technique called *independence detection*.

In this section, we will describe the original method of independence detection proposed by Standley (2010). The main idea behind this technique is that difficulty of MAPF solving optimally grows exponentially with the number of agents. It would be ideal, if we could divide the problem into a series of smaller sub problems, solve them independently at low computational effort, and then combine them.

The simple approach, called *simple independence detection* (SID), assigns each agent to a group so that every group consists of exactly one agent. Then, for each of these groups, an optimal solution is found independently. Every pair of these solutions is evaluated and if the two groups' solutions are in conflict (that is, when collision of agents belonging to different group occurs), the groups are merged and replanned together. If there are no conflicting solutions, the solutions can be merged to a single solution of the original problem. This approach can be further improved by avoiding merging of groups.

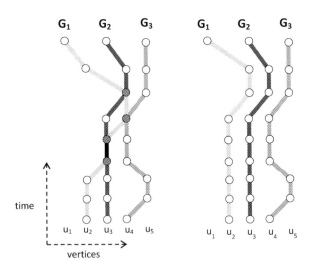

Fig. 4. A schematic illustration from [31] of path replanning within the independence detection technique. A path for the group G_1 conflicted with paths of other two groups (left part). Then path for G_1 has been successfully replanned (right part).

Generally, each agent has more than one possible optimal path. However, SID considers only one of these paths. The improvement of SID known as independence detection (ID) is as follows. Let's have two conflicting groups G_1 and G_2. First, try to replan G_1 so that the new solution has the same cost and the steps that are in conflict with G_2 are forbidden.

If no such solution is possible, try to similarly replan G_2. If this is not possible, merge G_1 and G_2 into a new group. In case either of the replanning was successful, that group needs to be evaluated with every other group again. This can lead to infinite cycle. Therefore, if two groups were already in conflict before, merge them without trying to replan.

Algorithm 2. MAPF solving algorithm based on **independence detection** (ID) technique. Planning for groups is always done to have least number of conflicts w.r.t. conflict avoidance table.

```
assign each agent to a group;
plan a path for each group by A*;
fill conflict avoidance table;
while conflicting groups exist
  G₁, G₂ = conflicting groups;
  if G₁, G₂ not conflicted before
    replan G₁ by A* with illegal moves based on G₂;
    if failed to replan G₁
      replan G₂ by A* with illegal moves based on G₁;
    endif
  endif
  if no alternate paths for G₁, G₂
    merge G₁ and G₂;
    plan a path for new group by A*;
  endif
  update conflict avoidance table;
end
return combined paths of all groups;
```

Standley uses ID in combination with the A* algorithm. While planning, it is preferred to find paths that create the least possible amount of conflicts with other groups that have already planned paths. For this purpose, the conflict avoidance table is created (see Algorithm 2 for pseudo-code).

The table stores moves of agents in other groups. In case A* has a choice between several nodes with the same minimal f() cost, the one with least amount of conflicts is expanded first. This technique yields an optimal solution that has a minimal number of conflicts with other groups. This property is useful when replanning of a group's solution is needed.

Both SID and ID do not solve MAPF on their own, they only divide the problem into smaller sub-problems that are solved by any possible MAPF algorithms. Thus, ID and SID are general frameworks, which can be executed on top of any MAPF solver.

6   Integrating SID and ID into MDD-SAT

SID can be integrated into the SAT-based framework as a top-level algorithm where MDD-SAT merely serves as a procedure for optimal MAPF solving restricted on an individual group. Hence, no modification of the core MDD-SAT procedure is needed.

ID however requires modification of the original ID since in the propositional formula it is not possible to express preference that individual paths of groups of agents should avoid occupied positions in the *conflict avoidance table*. In the yes/no SAT environment we either manage to avoid occupied positions or not while in the negative case there is no easy tool how to control the number of conflicts.

The SAT-based version of ID works in similar way to the original version of Standley but instead of resolving conflicts between a pair of conflicting groups G_1 and G_2 it resolves conflict of group G_1 with all other groups. If this attempt is successful, G_1 is independent on others and the process can continue with resolving conflicts between remaining groups (see Fig. 4 where G_1 has been made independent).

If the attempt to resolve conflict between G_1 and G_2 by making G_1 independent fails, the same is tried for G_2. If the attempt for G_2 fails too groups are merged. The pseudo-code is shown as Algorithm 3.

In contrast to original ID we strictly require avoidance with respect to the conflict avoidance table instead of stating it as a preference only. This is technically done by omitting the conflicting vertices in the MDD. The SAT approach does not allow to express a preference like in the search based algorithm. This is the reason why ID in the SAT-based solver differs from the original one.

Algorithm 3. Independence detection in the **SAT-based framework**. Conflict aviodance is strictly required.

```
assign each agent to a group;
plan a path for each group G₁,…,Gₖ by MDD-SAT;
fill conflict avoidance table;
while conflicting groups exist
   G₁, G₂ = conflicting groups;
   if G₁, G₂ not conflicted before
      replan G₁ by MDD-SAT with illegal moves based on
      {G₁,…,Gₖ}-G₁;
      if failed to replan G₁
         replan G₂ by MDD-SAT with illegal moves based on
         {G₁,…,Gₖ}-G₂;
      endif
   endif
   if no alternate paths for G₁, G₂
      merge G₁ and G₂;
      plan a path for new group by MDD-SAT;
   endif
   update conflict avoidance table;
end
return combined paths of all groups;
```

7   Experiments

We performed experimental comparison of the proposed MDD-SAT+SID and MDD-SAT+ID solvers with other state-of-the-art solvers – namely with the previous best SAT-based solver MDD-SAT and also with search-based algorithms ICTS and ICBS.

The MDD-SAT+SID and MDD-SAT+ID have been implemented in C++ as an extension of an existing implementation of the MDD-SAT solver. A couple of minor improvements have been done in the original MDD-SAT encoding – some auxiliary propositional variables have been eliminated which reduced the size of the encoding and consequently saved runtime while generating formulae (this improvement affects both MDD-SAT and new MDD-SAT+SID, MDD-SAT+ID used in presented experiments).

We used Glucose 3.0 [1] in variants of MDD-SAT which is a top performing SAT solver according to the recent SAT Competitions [4]. The complete implementation of the MDD-SAT solvers is available on-line to allow reproducibility of the presented results: http://ktiml.mff.cuni.cz/~surynek/research/icaart2017.

ICTS and ICBS have been implemented in C#. The original implementations of these algorithms have been used.

All the tests were run on Xeon 2 Ghz, and on Phenom II 3.6 Ghz, both with 12 Gb of memory.

The experimental setup followed the scheme used in the literature [22] which tests MAPF algorithms on 4-connected grids. Let us note however that all the suggested algorithms are designed and implemented for general undirected graphs (the fact that grids are used in the experiments is not exploited to increase efficiency of solving in any way).

7.1   Small Grids Evaluation

The first series of experiments takes place on small square grids of sizes 8×8, 16×16, and 32×32 with 10% of vertices occupied by obstacles. In this setup of the environment, we increased population of agents from 1 and observed the runtime of all the solvers until no solver was able to solve the instance within the given time limit of 300 s (this was 20 agents for 8×8 grid, and 40 and 60 for 16×16 and 32×32 girds respectively).

Ten randomly generated instances per number of agents were used. The initial positions were generated by choosing a subset of vertices randomly. The goal arrangement has been generated as a long random walk from the initial state following valid moves – this ensured solvability of all the tested instances.

To be able to communicate results of experiments more easily we intuitively distinguish three different categories of instances with respect to the density of agents as follows. The behavior of solvers is then discussed with respect to these categories:

- **Low density** – few interactions among agents, paths for individual agents can be planned independently.
- **Medium density** – some interaction among agents are inevitable but there exist multiple groups of agents that are independent of each other.
- **High density** – majority of agents are interdependent and form one large group.

The small grid experiment contains instances from all these three cases. The hypothesis is that the SID and ID technique will be helpful in instances with medium density of agents while ID is expected to reach benefit in higher densities of agents. We also expect that in the case of low density of agents there will be some benefit of SID and ID since many agents will just follow their shortest paths towards goals in such a case. As in low and medium density cases the complexity of the formula is not proportional to the difficulty of the instance.

Furthermore, we expect rather negative effect of using SID and ID in instances with high density of agents. This is because of the fact that most agents will be gradually merged into a large group while the process of merging represents an overhead in such a case.

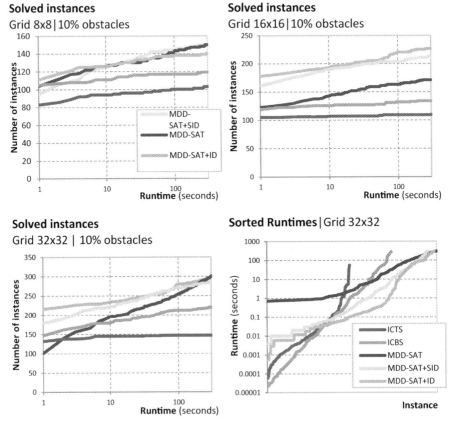

Fig. 5. Results of experiments on **small grid maps** of sizes 8×8, 16×16, and 32×32. Figures show how many instances were solved within the given runtime and sorted runtimes (right bottom part). Clearly versions of MDD-SAT dominate in the test over search based algorithms ICTS and ICBS except few quickly solvable cases. Moreover, MDD-SAT+ID and MDD-SAT+SID outperforms MDD-SAT in cases with low to medium density of agents. MDD-SAT+ID and MDD-SAT+SID exhibit similar performance while ID shows its advantage in instances requiring more time.

Experimental result for the small grids (see Fig. 5) confirmed the hypothesis. MDD-SAT+SID/ID win in low to medium density of agents. For the higher density of agents, both MDD-SAT+SID/ID tend to be eventually outperformed by the original MDD-SAT. If SID and ID are compared then we can see that ID has more significant benefit than SID in most cases.

7.2   Large Maps – Dragon Age

We also experimented on three structurally different large maps from Dragon Age: Origins [26] – ost003d, den520d, and brc202d (see Fig. 6). Our choice of maps is driven by the choice of authors in the previous literature [20, 30].

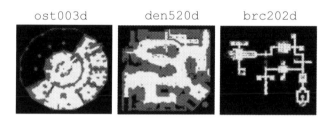

Fig. 6. Illustration of large Dragon Age maps ost003d (size 194 × 194), den520d (size 257 × 256), and brc202d (size 481 × 530).

We used setup with 16 and 32 agents randomly paced agents which represents low to medium density. Let us note that a case with high density of agents in the map of that size is currently out of reach of any existing algorithm.

To obtain problems of various difficulties the distance of agents from initial positions to their goals has been varied in the range 8, 16, 24, ..., 320.

For each distance 10 random instances were generated in which initial positions were selected randomly and then random walk has been performed until all the agents reach at least the given distance from its initial position.

The hypothesis for large maps is that MDD-SAT+SID/ID should dominate generally with some expected advantage of ID which in fact is the same hypothesis as in the case of small grids because here we have only the low-medium density case. However, as there are important structural differences between the three tested maps which impact is hardly predictable. Intuitively, SID/ID should have been more beneficial in ost003d and den520d maps since in these maps there is more room to find alternative paths.

Results for the three Dragon Age maps are shown in Figs. 7, 8, and 9. Again the number of instances solved in the given runtime is shown. The difficulty (runtime) grows with the growing distance of agents from their goals in this setup.

It can be read from these results that MDD-SAT+ID tends to outperform MDD-SAT in more difficult instances. In these instances, the interaction among agents in nontrivial but on the other hand the interdependence among agents is tractable by ID.

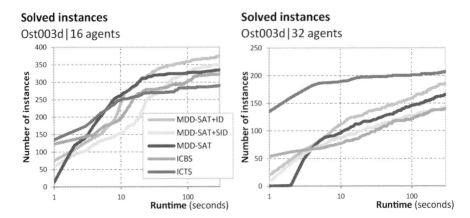

Fig. 7. Results of experiments on Dragon Age map ost003d. MDD-SAT+ID outperforms MDD-SAT in harder instances while MDD-SAT+SID performs worse than MDD-SAT. All MDD-SAT versions are dominated by ICTS.

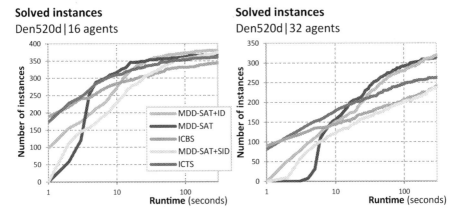

Fig. 8. Results of experiments on Dragon Age map den520d. ID brings minor benefit in harder instances while SID has merely a negative effect.

Surprising results have been obtained for MDD-SAT+SID which performed generally worse than MDD-SAT. SID hence was unsuccessful in independence detection enough to produce any performance benefit in MDD-SAT expect the case of very easy instances.

The intuitive hypothesis was not confirmed completely since surprisingly MDD-SAT is better than MDD-SAT+ID in easier instances of medium density category usually and the performance of MDD-SAT+SID remained behind expectation. Our initial intuitive hypothesis did estimate well the effort needed for merging groups that eventually represents a big overhead in case of large maps. Hence, MDD-SAT+ID can

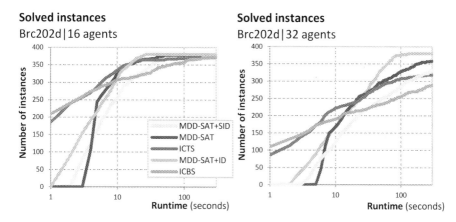

Fig. 9. Results of experiments on Dragon Age map `brc202d`. ID brings significant improvement in harder instances with 32 agents. SID again has rather a negative effect in MDD-SAT.

show its benefit after the difficulty of the formula representing the entire MAPF instance prevails over the difficulty of group merging.

Another surprising result was obtained in `brc202d` map where MDD-SAT+ID was a very clear winner in harder instances with 32 agents.

Moreover, we cannot say that SAT-based approach represented by MDD-SAT and MDD-SAT+SID/ID is a universal winner as there are cases where ICTS and ICBS dominate (`ost003d` with 32 agents is such an example).

7.3   Discussion

It can be generally observed that ID brings worthwhile improvement to MDD-SAT solver which by itself performs very well. The simple version of independence detection SID provides worse benefit than ID and in large instances its effect is even negative.

Experimental results indicate that there is a certain range of the density of agents though not precisely determined in our evaluation in which ID is beneficial while outside this range it cases an overhead.

The implementation of ID within the MDD-SAT+ID solver did not use any special reasoning about what groups of agents should be merged or not. The groups were processed in the ordering given by the original ordering of agents. We expect that more careful reasoning about merging can bring yet more improvements.

8   Conclusion

Integration of the existing technique of independence detection (ID) into the SAT-based approach to MAPF has been presented. Performed experimental evaluation shows that using ID in the SAT-based approach to MAPF has a significant performance benefit. This can be especially observed in instances with medium density of agents where interactions among agents are non-trivial but there is still chance to find sufficiently many small independent groups of agents.

The new solver called MDD-SAT+ID is a combination of an existing SAT-based MAPF solver MDD-SAT and ID. We have shown that MDD-SAT+ID is a new state-of-the-art in the optimal SAT-based MAPF solving for the presented categories of the MAPF problem. Moreover, the new MDD-SAT+ID performs well also with respect to the best search based solvers like ICTS and ICBS though we cannot say that MDD-SAT+ID is a universal winner.

The future research of the presented topic will focus on the following aspects: (i) The classification of density of agents was intuitive only in the presented experimental evaluation. Hence the immediate future work is to develop concepts for more precise classification of the density and interaction among agents. (ii) ID is not beneficial across all instances and sometimes represents an overhead. Hence, having a more rigorous classification of the density of agents, we can develop a mechanism for automated deciding whether to use ID or not according to the classified density of agents. (iii) Currently we take groups of agents to be merged in the same order as they appear in the input. A more careful consideration of which groups to merge may lead to further performance improvements.

Acknowledgements. This paper is supported by the joint grant of the Israel Ministry of Science and the Czech Ministry of Education Youth and Sports number 8G15027, and Charles University under the SVV project number 260 333.

We would like thank anonymous reviewers for their constructive comments of [31] which helped us to prepare this extended version of the paper.

References

1. Audemard, G., Simon, L.: The Glucose SAT Solver (2013). http://labri.fr/perso/lsimon/glucose/. Accessed Oct 2016
2. Audemard, G., Simon, L.: Predicting learnt clauses quality in modern SAT solvers. In: Proceedings of the 21st International Joint Conference on Artificial Intelligence (IJCAI 2009), pp. 399–404. IJCAI (2009)
3. Bailleux, O., Boufkhad, Y.: Efficient CNF encoding of boolean cardinality constraints. In: Rossi, F. (ed.) CP 2003. LNCS, vol. 2833, pp. 108–122. Springer, Heidelberg (2003). https://doi.org/10.1007/978-3-540-45193-8_8
4. Balint, A., Belov, A., Heule, M., Järvisalo, M.: SAT 2015 competition (2015). http://www.satcompetition.org/. Accessed Oct 2016
5. van den Berg, J., Snoeyink, J., Lin, M.C., Manocha, D.: Centralized path planning for multiple robots: optimal decoupling into sequential plans. In: Proceedings of Robotics: Science and Systems V, University of Washington. The MIT Press (2010)
6. Biere, A., Heule, M., van Maaren, H., Walsh, T.: Handbook of Satisfiability. IOS Press, Amsterdam (2009)

7. Boyarski, E., Felner, A., Stern, R., Sharon, G., Tolpin, D., Betzalel, O., Shimony, S.: ICBS: improved conflict-based search algorithm for multi-agent pathfinding. In: Proceedings of the 24th International Joint Conference on Artificial Intelligence (IJCAI 2015), pp. 740–746. IJCAI (2015)

8. Čáp, M., Novák, P., Vokřínek, J., Pěchouček, M.: Multi-agent RRT: sampling-based cooperative pathfinding. In: International conference on Autonomous Agents and Multi-Agent Systems (AAMAS 2013), pp. 1263–1264. IFAAMAS (2013)

9. Erdem, E., Kisa, D.G., Öztok, U., Schüller, P.: A general formal framework for pathfinding problems with multiple agents. In: Proceedings of the 27th AAAI Conference on Artificial Intelligence (AAAI 2013). AAAI Press (2013)

10. Huang, R., Chen, Y., Zhang, W.: A novel transition based encoding scheme for planning as satisfiability. In: Proceedings of the 24th AAAI Conference on Artificial Intelligence (AAAI 2010). AAAI Press (2010)

11. Kautz, H., Selman, B.: Unifying SAT-based and graph-based planning. In: Proceedings of the 16th International Joint Conference on Artificial Intelligence (IJCAI 1999), pp. 318–325. Morgan Kaufmann (1999)

12. Kim, D., Hirayama, K., Park, G.-K.: Collision avoidance in multiple-ship situations by distributed local search. J. Adv. Comput. Intell. Intell. Inform. (JACIII) 18(5), 839–848 (2014)

13. Kornhauser, D., Miller, G.L., Spirakis, P.G.: Coordinating pebble motion on graphs, the diameter of permutation groups, and applications. In: Proceedings of the 25th Annual Symposium on Foundations of Computer Science (FOCS 1984), pp. 241–250. IEEE Press (1984)

14. Ma, H., Koenig, S., Ayanian, N., Cohen, L., Hoenig W., Kumar, T.K.S., Uras, T., Xu, H., Tovey, C., Sharon, G.: Overview: generalizations of multi-agent path finding to real-world scenarios. In: IJCAI-16 Workshop on Multi-Agent Path Finding (WOMPF) (2016)

15. Michael, N., Fink, J., Kumar, V.: Cooperative manipulation and transportation with aerial robots. Auton. Robot. 30(1), 73–86 (2011)

16. Ratner, D., Warmuth, M.K.: NxN puzzle and related relocation problems. J. Symb. Comput. 10(2), 111–138 (1990)

17. Ryan, M.R.K.: Exploiting Subgraph structure in multi-robot path planning. J. Artif. Intell. Res. (JAIR) 31, 497–542 (2008)

18. Ryan, M.R.K.: Constraint-based multi-robot path planning. In: Proceedings ICRA 2010, pp. 922–928. IEEE Press (2010)

19. Sharon, G., Stern, R., Goldenberg, M., Felner, A.: The increasing cost tree search for optimal multi-agent pathfinding. Artif. Intell. 195, 470–495 (2013)

20. Sharon, G., Stern, R., Felner, A., Sturtevant, N.R.: Conflict-based search for optimal multi-agent pathfinding. Artif. Intell. 219, 40–66 (2015)

21. Marques-Silva, J., Lynce, I.: Towards robust CNF encodings of cardinality constraints. In: Bessière, C. (ed.) CP 2007. LNCS, vol. 4741, pp. 483–497. Springer, Heidelberg (2007). https://doi.org/10.1007/978-3-540-74970-7_35

22. Silver, D.: Cooperative pathfinding. In: Proceedings of the 1st Artificial Intelligence and Interactive Digital Entertainment Conference (AIIDE 2005), pp. 117–122. AAAI Press (2005)

23. Sinz, C.: Towards an optimal CNF encoding of boolean cardinality constraints. In: van Beek, P. (ed.) CP 2005. LNCS, vol. 3709, pp. 827–831. Springer, Heidelberg (2005). https://doi.org/10.1007/11564751_73

24. Standley, T.: Finding optimal solutions to cooperative pathfinding problems. In: Proceedings of the 24th AAAI Conference on Artificial Intelligence (AAAI 2010), pp. 173–178. AAAI Press (2010)

25. Standley, T., Korf, R.E.: Complete algorithms for cooperative pathfinding problems. In: Proceedings of the 22nd International Joint Conference on Artificial Intelligence (IJCAI 2011), pp. 668–673. IJCAI (2011)
26. Sturtevant, N.R.: Benchmarks for grid-based pathfinding. IEEE Trans. Comput. Intell. AI Games **4**(2), 144–148 (2012)
27. Surynek, P.: A novel approach to path planning for multiple robots in biconnected graphs. In: Proceedings of the 2009 IEEE International Conference on Robotics and Automation (ICRA 2009), pp. 3613–3619. IEEE Press (2009)
28. Surynek, P.: An optimization variant of multi-robot path planning is intractable. In: Proceedings of the 24th AAAI Conference on Artificial Intelligence (AAAI 2010), pp. 1261–1263. AAAI Press (2010)
29. Surynek, P.: Compact representations of cooperative path-finding as SAT based on matchings in bipartite graphs. In: Proceedings of the 26th IEEE International Conference on Tools with Artificial Intelligence (ICTAI 2014), pp. 875–882. IEEE Computer Society (2014)
30. Surynek, P., Felner, A., Stern, R., Boyarski, E.: Efficient SAT approach to multi-agent path finding under the sum of costs objective. In: Proceedings of 22nd European Conference on Artificial Intelligence (ECAI 2016), pp. 810–818. IOS Press (2016)
31. Surynek, P., Švancara, J., Felner, A., Boyarski, E.: Integration of independence detection into SAT-based optimal multi-agent path finding: a novel SAT-based optimal MAPF solver. In: Proceedings of the 9th International Conference on Agents and Artificial Intelligence (ICAART 2017). SciTe Press (2017)
32. Yu, J., LaValle, S.M.: Structure and intractability of optimal multirobot path planning on graphs. In: Proceedings of the 27th AAAI Conference on Artificial Intelligence (AAAI 2013). AAAI Press (2013)
33. Yu, J., LaValle, S.M.: Planning optimal paths for multiple robots on graphs. In: Proceedings of the IEEE International Conference on Robotics and Automation (ICRA 2013), pp. 3612–3617. IEEE Press (2013)
34. Wang, K.C., Botea, A.: Fast and memory-efficient multi-agent pathfinding. In: Proceedings of the 18th International Conference on Automated Planning and Scheduling (ICAPS 2008), pp. 380–387. AAAI Press (2008)
35. de Wilde, B., ter Mors, A., Witteveen, C.: Push and rotate: a complete multi-robot pathfinding algorithm. J. Artif. Intell. Res. (JAIR) **51**, 443–492 (2014)
36. Wilson, R.M.: Graph puzzles, homotopy, and the alternating group. J. Comb. Theory Ser. B **16**, 86–96 (1974)

ε-Strong Privacy Preserving Multi-agent Planning

Antonín Komenda$^{(\boxtimes)}$, Jan Tožička, and Michal Štolba

Department of Computer Science, Faculty of Electrical Engineering,
Czech Technical University in Prague,
Karlovo náměstí 13, 121 35 Prague, Czech Republic
{antonin.komenda,jan.tozicka,michal.stolba}@fel.cvut.cz

Abstract. Multi-agent planning can solve various sequential decision problems comprising multiple entities. In contrast to classical planning, the agents are interested in maintaining privacy while planning with each other. Therefore they have to reason about what information they can share. Although privacy is one of the crucial aspects of multi-agent planning, formal and algorithmic treatment of privacy is rather sparse in literature. No domain-independent strong privacy preserving multi-agent planner was proposed so far. Moreover, our recent results indicate that an efficient variant of such planner may not exist at all. Such strong privacy preserving planner would not allow to leak any private information during planning neither directly nor indirectly. Especially the indirect leakage is hard to assess as it can be based on any possible deduction principle from the non-private information along the planning process.

Here, we propose a refined version of a multi-agent planning principle, based on our previous work published as the conference version of this paper. The planning principle is designed so that it can get arbitrarily close to the general strong privacy preserving planning for the price of decreased planning efficiency. We have tighten the bounds on the privacy leakage and proved the strong privacy can be achieved by a finite number of additional plans, in contrast to the previous algorithm, where the number had to be infinite in general. We newly illustrate the principle on an additional synthetic planning problem, which shows the general privacy leakage upper bound. As in the previous variant of the algorithm, the strong privacy assurances are under computational tractability assumptions commonly used in secure computation research.

Keywords: Automated planning · Multi-agent systems · Privacy
Security

1 Introduction

Multi-agent planning deals with a problem of finding a coordinated sequence of actions for a set of entities (or agents), such that the actions are applicable from

© Springer International Publishing AG, part of Springer Nature 2018
J. van den Herik et al. (Eds.): ICAART 2017, LNAI 10839, pp. 137–156, 2018.
https://doi.org/10.1007/978-3-319-93581-2_8

a predefined initial state and transform the environment to a state where predefined goals are fulfilled. If the environment and the actions are deterministic (that is their outcome is unambiguously defined by the state they are applied in), the problem is a deterministic multi-agent planning problem [3]. Furthermore, if the set of goals is common to all agents and the agents cooperate in order to achieve the goals, the problem is a cooperative multi-agent planning problem. The reason the agents cannot simply feed their problem descriptions into a centralized planner typically lies in that although the agents cooperate, they want to share only the information necessary for their cooperation, but not the information about their inner processes. Such privacy constraints are respected by privacy preserving multi-agent planners.

A number of privacy preserving multi-agent planners have been proposed in recent years, such as MAFS [10], FMAP [15], MADLA [19], PSM [16] and GPPP [8]. Although all of the mentioned planners claim to be privacy-preserving, formal proofs of such claims do not exist. The privacy of MAFS is discussed in [10] and expanded upon in [2], proposing Secure-MAFS, a version of MAFS with stronger privacy guarantees. These guarantees are proven for a family of planning problems, but does not hold generally. The approach was recently generalized in the form of Macro-MAFS [7], however without strengthening the claims about privacy.

Here, we propose a parameterized variant of strong privacy preserving planning using the definition of privacy in [10]. We show how the two extremities of the parameter lead to strong privacy preserving, but inefficient planner; or weak privacy preserving, but efficient planner.

This article is a reworked and extended version of the paper [17]. In this version, we have reformulated most of the definitions and proofs to improve readability and conciseness of the arguments. We provide tighter bounds on privacy leakage and propose a novel example illustrating the principle of the proposed planner. Moreover, the novel example provides a ground for novel claim about privacy leakage in multi-agent planning in general.

2   Multi-agent Planning

The most common model for multi-agent planning is MA-STRIPS [3] and derived models (such as MA-MPT [10] using multi-valued variables). We reformulate the MA-STRIPS definition and we also generalize the definition to multi-valued variables. Formally, for a set of agents $\mathcal{A}$, a problem $\mathcal{M} = \{\Pi_i\}_{i=1}^{|\mathcal{A}|}$ is a set of agent problems. An agent problem of agent $\alpha_i \in \mathcal{A}$ is defined as

$$\Pi_i = \left\langle \mathcal{V}_i = \mathcal{V}^{\text{pub}} \cup \mathcal{V}_i^{\text{priv}}, O_i = O_i^{\text{pub}} \cup O_i^{\text{priv}} \cup O^{\text{proj}}, s_I, s_\star \right\rangle,$$

where $\mathcal{V}_i$ is a set of variables s.t. each $V \in \mathcal{V}_i$ has a finite domain $\text{dom}(V)$, if all variables are binary (i.e. $|\text{dom}(V)| = 2$), the formalism corresponds to MA-STRIPS. The set of variables is partitioned into the set $\mathcal{V}^{\text{pub}}$ of public variables (with all values public), common to all agents and the set $\mathcal{V}_i^{\text{priv}}$ of variables private to α_i (with all values private), such that $\mathcal{V}^{\mathsf{pub}} \cap \mathcal{V}_i^{\mathsf{priv}} = \emptyset$. A complete assignment over $\mathcal{V}$ is a *state*, partial assignment over $\mathcal{V}$ is a *partial state*. We denote $s[V]$ as the value of V in a (partial) state s and $\mathsf{vars}(s)$ as the set of variables defined in s. The state s_I is the initial state of the agent α_i containing only $\mathcal{V}^{\mathsf{pub}}$ and $\mathcal{V}_i^{\mathsf{priv}}$ variable. $s_\star$ is a partial state representing the goal condition, that is if for all variables $V \in \mathsf{vars}(s_\star)$, $s_\star[V] = s[V]$, s is a goal state. Similarly, as in [9], we require all goals to be public, as private goals can be transformed into a public equivalent [16], i.e. $\mathsf{vars}(s_\star) \subseteq \mathcal{V}^{\mathsf{pub}}$.

The set O_i of actions comprises of a set O_i^{priv} of private actions of α_i, a set O_i^{pub} of public actions of α_i. A public projection of an action removes all its private parts. The set O^{proj} contain public projections of other agents' actions. O_i^{pub}, O_i^{priv}, and O^{proj} are pairwise disjoint. An action is defined as a tuple $a = \langle \mathsf{pre}(a), \mathsf{eff}(a) \rangle$, where $\mathsf{pre}(a)$ and $\mathsf{eff}(a)$ are partial states representing the precondition and effect respectively. An action a is applicable in a state s if $s[V] = \mathsf{pre}(a)[V]$ for all $V \in \mathsf{vars}(\mathsf{pre}(a))$ and the application of a in s, denoted $a \circ s$, results in a state s' s.t. $s'[V] = \mathsf{eff}(a)[V]$ if $V \in \mathsf{vars}(\mathsf{eff}(a))$ and $s'[V] = s[V]$ otherwise. A public action can be defined using a mixture of public and private preconditions and effects. A private action can be defined only over the private variables. As we often consider the planning problem from the perspective of agent α_i, we omit the index i.

We model all "other" agents as a single agent (the adversary), as all the agents can collude and combine their information in order to infer more. The public part of the problem Π which can be shared with the adversary is denoted as a public projection. The public projection of a (partial) state s is $s^{\triangleright}$, restricted only to variables in $\mathcal{V}^{\mathsf{pub}}$, that is $\mathsf{vars}(s^{\triangleright}) = \mathsf{vars}(s) \cap \mathcal{V}^{\mathsf{pub}}$. We say that s, s' are publicly equivalent states if $s^{\triangleright} = s'^{\triangleright}$. The public projection of action $a \in O^{\mathsf{pub}}$ is $a^{\triangleright} = \langle \mathsf{pre}(a)^{\triangleright}, \mathsf{eff}(a)^{\triangleright} \rangle$ and of action $a' \in O^{\mathsf{priv}}$ is an empty action noop. The public projection of Π is $\Pi^{\triangleright} = \langle \mathcal{V}^{\mathsf{pub}}, \{a^{\triangleright} | a \in O^{\mathsf{pub}}\}, s_I^{\triangleright}, s_\star^{\triangleright} \rangle$.

We define the solution to Π as follows. A sequence $\pi = (a_1, ..., a_k)$ of actions from O, s.t. a_1 is applicable in $s_I = s_0$ and for each $1 \leq i \leq k$, a_i is applicable in s_{i-1} and $s_i = a_i \circ s_{i-1}$, is a local s_k-plan, where s_k is the resulting state. If s_k is a goal state, π is a local plan, that is a local solution to Π. A local plan contains actions only of one agent, public, private or projected. Note that the actions in O^{proj} are assumed to be of the particular agent as well.

Such local plan π does not have to be the global solution to $\mathcal{M}$, as the actions of other agents (O^{proj}) are used only as public projections and are missing private preconditions and effects of other agents. The public projection of π is defined as $\pi^{\triangleright} = (a_1^{\triangleright}, ..., a_k^{\triangleright})$ with the noop actions omitted.

From the global perspective of $\mathcal{M}$ a public plan $\pi^{\triangleright} = (a_1^{\triangleright}, ..., a_k^{\triangleright})$ is a sequence of public projections of actions of various agents from $\mathcal{A}$ such that the actions are sequentially applicable with respect to $\mathcal{V}^{\mathsf{pub}}$ starting in $s_I^{\triangleright}$ and the resulting state satisfies $s_\star^{\triangleright}$. A public plan is α_i-extensible, if by replacing $a_{k'}^{\triangleright}$ s.t. $a_{k'} \in O_i^{\mathsf{pub}}$ by the respective $a_{k'}$ and adding $a_{k''} \in O^{\mathsf{priv}}$ to required places we obtain a local plan (solution) to Π_i. According to [16], a public plan $\pi^{\triangleright}$ α_i-extensible by all $\alpha_i \in \mathcal{A}$ is a global solution to $\mathcal{M}$.

The nature of MA-MPT planning allows for plans containing repeated action sequences (in extreme, repeated infinitely many times). Such repetitions however does not provide transformation to a state not yet visited. Therefore the length of global *meaningful* plans is bounded by the number of all possible states

$$\prod_{V \in \mathcal{V}^{\mathsf{pub}} \cup \bigcup_{i=1}^{|\mathcal{A}|} \mathcal{V}_i^{\mathsf{priv}}} |\mathsf{dom}(V)|. \tag{1}$$

We define the sets of all meaningful global and local plans as $\text{SOLS}(\mathcal{M})$ and $\text{SOLS}(\Pi)$ respectively. Lengths of plans in both sets are limited by the presented bound on meaningful plans and therefore the sets are finite.

Since an agent can be required to repeatedly produce a particular value assignment of a variable, which is consumed (needed in precondition and changed in effect) by an action of another agent, we have to allow for meaningful repetitions in $\text{SOLS}(\Pi)$. Length of a local plan with such repetitions is however still limited by the maximal length of its related global plan solving $\mathcal{M}$. Therefore we use the same bound on length both for plans in $\text{SOLS}(\mathcal{M})$ and $\text{SOLS}(\Pi)$ where $\Pi \in \mathcal{M}$. $\text{SOLS}_l(\Pi)$ will denote sets of local plans of length l; $\text{SOLS}_{\leq l}(\Pi)$ and $\text{SOLS}_{>l}(\Pi)$ denote sets of local plans of length no more than l and longer than l respectively. Finally, $\text{SEQS}(\Pi)$ will denote all possible sequences of actions (incl. non-plans) of the problem Π.

2.1 Privacy

First definition of privacy leakage quantification was proposed in [18]. It was based on enumeration of all plans, which we used as the underlying principle for measuring the privacy leakage in this work. However, we do not explicitly require enumeration of the particular plans as in looser form, our bounds work with all possible action sequences. Note that the work in [18] is also not easily applicable to MA-STRIPS and MA-MPT.

The only rigorous definition of privacy for MA-STRIPS and MA-MPT so far was proposed in [10] and extended in [2]. The authors present two notions, weak and strong privacy preservation:

An algorithm is *weak privacy-preserving* if, during the whole run of the algorithm, the agent does not openly communicate private parts of the states, private actions and private parts of the public actions. In other words, the agent openly communicates only the information in $\Pi^{\triangleright}$. Even if not communicated, the adversary may deduce the existence and values of private variables, preconditions and effects from the (public) information communicated.

An algorithm is *strong privacy-preserving* if the adversary can deduce no information about a private variable and its values and private preconditions/effect of an action, beyond what can be deduced from the public projection $\Pi^{\triangleright}$ and the public projection of the solution plan $\pi^{\triangleright}$.

2.2   Secure Computation

In general, any function can be computed securely [1,20,21], however it is not known how to encode the whole planning process into one function [14]. In this contribution, we focus on more narrow problem of *private set intersection* (PSI), where each agent has a private set of numbers and they want to securely compute the intersection of their private sets while not disclosing any numbers which are not in the intersection. The *ideal PSI* supposes that no information is transferred between the agents [11].

Ideal PSI can be solved with trusted third party which receives both private sets, computes the intersection, and sends it back to agent. As long as the third party is honest, the computation is correct and no information leaks.

In literature (e.g., [6,11]), we can find several approaches how the ideal PSI can be solved without trusted third party. Presented solutions are based on several computational hardness assumptions, e.g., intractable large number factorization, DiffieHellman assumption [4], etc. All these assumptions break when an agent has access to unlimited computation power, therefore all the results hold under the assumption that $P \neq NP$, in other words by computational intractability of breaking PSI.

3   ε-Strong Privacy Preserving Multi-agent Planner

Multi-agent planner fulfilling the strong privacy requirement forms the lower bound of information exchanged between the agents. Agents do not leak any information about their internal problems and thus their cooperation cannot be effective [14], nevertheless, a strong privacy preserving multi-agent planner is an important theoretical result that could lead to better understanding of privacy preservation during multi-agent planning and consequentially also to creation of more privacy preserving planners.

In this contribution, we present a planner that is not strong privacy preserving but can be arbitrarily close to it. We focus on planning using projected plan-space[1] search [13] and thus we will define the terms in that respect. In the following definitions and proofs we suppose that there are two semi-honest (honest but curious) agents α_- and α_+. We will consider the perspective of the agent α_- trying to detect the private information of α_+ for the simplicity of the presentation, but all holds for both agents and also for a larger group of agents. Similarly to [2], we also assume that $O^{priv} = \emptyset$. This assumption can be stated WLOG as each sequence of private actions followed by a public action can be compiled to a single public action.

Definition 1 (Public Plan Acceptance). Public plan acceptance $P(\pi^{\triangleright})$ is a probability known to agent α_- whether a plan $\pi^{\triangleright}$ is α_+-extensible.

When the algorithm starts, α_- has some *a priori* information $P^0(\pi^{\triangleright})$ about acceptance of plan $\pi^{\triangleright}$ by agent α_+ (e.g., 0.5 probability of acceptance of each

---

[1] Projected plan-space contains all the solutions of the projected public problem $\Pi^{\triangleright}$.

plan in the case when α_- knows nothing about α_+). At the end of execution of the algorithm, this information changes to $P^*(\pi^\triangleright)$. Obviously, every agent knows that the solution public plan π^* the agents agreed on is extensible and thus it is accepted by every agent, i.e. $P^*(\pi^*) = 1$. The difference between the α_-'s a priori information and the final information represents information which leaked from α_+ during their communication. Whether an agent is *certain* about acceptance of a plan can be expressed as $|1 - 2P(\pi^\triangleright)|$, normalized to an interval $\langle 0, 1 \rangle$, where 0 means not knowing anything about acceptance of the plan $(P(\pi^\triangleright) = 0.5)$ and 1 means certainty $(P(\pi^\triangleright) = 1$ or $P(\pi^\triangleright) = 0)$.

Definition 2 (Leaked Information). Leaked information *from perspective of one agent during execution of a multi-agent planner leading to a solution π^* is a sum of changes in certainty about acceptance of the plan from the beginning $P^0(\pi^\triangleright)$ to the end $P^*(\pi^\triangleright)$ of planning excluding the solution plan π^**

$$\lambda = \sum_{\pi^\triangleright \in \{\pi^\triangleright | \pi \in \text{SOLS}(\Pi)\} \setminus \{\pi^*\}} |1 - 2P^*(\pi^\triangleright)| - |1 - 2P^0(\pi^\triangleright)|. \qquad (2)$$

As we do not assume the agents intentionally increase uncertainty about acceptance of other agents by sending invalid plans (the honest but curious agents), the certainty about acceptance of a plan can only grow, i.e.

$$|1 - 2P^*(\pi^\triangleright)| - |1 - 2P^0(\pi^\triangleright)| \geq 0, \text{ thus } \lambda \geq 0.$$

Definition of algorithm's *leaked information* allows us to formally define *strong privacy* of a projected plan-space planning algorithm. sec:strong-priv-pres

Proposition 1. (Strong Privacy). *A planning algorithm is* strong privacy preserving *if it assures $\lambda = 0$.*

Proof. Any information leakage allowing deduction of private information (preconditions or effects) in agent's planning problem during planning affect probability of acceptance or rejection of plan projections by other agents as the preconditions and effects are the only principle preventing of acceptance or rejection of a sequence of actions. Therefore $\lambda = 0$ holds if and only if no information by Definition 2 leaked.

Definition 3 (ε-Strong Privacy Preserving Planner). *For given $\varepsilon > 0$, an planning algorithm is ε-strong privacy preserving if it leaks acceptance or rejection of no more than ε local plans, i.e. $\lambda \leq \varepsilon$.*

The high-level idea of our proposed planner (Algorithm 1) is based on a systematic *generate-and-test* principle, similar to our recent principle proposed in [16]. Local plans π are generated in parallel by all agents and their public projections $\pi^\triangleright$ are distributively tested whether there are some projections π^* common to all agents. Since only acceptable local plans π thus public projections $\pi^\triangleright$ are generated and tested, if a public projection common to all agents is found, it is guaranteed to be a global solution [16]. Provided that the distributed testing is done such that no information leaks, the only other point of possible information leakage is from the fact that a global solution was not found for a particular set of generated local plans. In other words, knowing α_+ refused all possible solutions in a well defined set of plans, tells α_- the plans were refused because of some private preconditions of α_+. Technically, the only parts of the algorithm, where the agents can learn something about each others' plans is therefore at lines 9 and 11, where all agents know that a solution was not found for all plans generated by the iterations of the algorithm so far.

Let us assume the systematicity of the generate-and-test principle is in testing of incrementally longer plans. The length l of the agents' local plans grows with each iteration of the main loop (lines 3, 4 and 13), therefore each distributed intersection (line 9) is done for generated plans of length $\leq l$. After each iteration, which did not end at line 11, the agent α_- knows that the agent α_+ refuses a local plan projection $\pi^{\triangleright}$ generated by α_-. This increases the certainty about refusal of $\pi^{\triangleright}$ and therefore increases λ. Note that such situation can be caused only by unfulfilled private preconditions of an action of the agent α_+ which prevent α_+ to generate $\pi^{\triangleright}$. This principle is known as privately dependent actions, for more detail see [12].

The principle of iterations synchronized by length was used only for the sake of clearer explanation. The argument however holds WLOG also for other iterative schemes. If the length of the generated plans is not synchronized by the iterations, all local plans of length l will be eventually generated by both agents α_+ and α_-. When a solution of length $l + 1$ is found, α_- can use the same reasoning as in the previous paragraph to deduce α_+ has some unfulfilled private preconditions in $\pi^{\triangleright}$.

Generally, the same line of reasoning can be even used for any systematic plan generating algorithm, under the assumption all agents know the other agents' systematic plan generation algorithms. There always exists a point in future when α_- knows that α_+ had generated a plan, which would have be a solution of the problem and the algorithm would have end. And the only reason, why this had happened is that α_+ has some unfulfilled private preconditions.

To fulfill the ε-privacy requirement by the Algorithm 1, the systematically generated plans, which can leak information, are supplemented by a sufficiently large amount of randomly generated plans on line 7. As these plans are longer than the systematic ones, with a probability proportional to the number of such plans generated, they can shortcut finding of a solution sooner than using only the systematic plan generation. The formula $|\text{SOLS}_{>l}(\Pi_i)|(1 - \frac{\varepsilon}{|\text{SOLS}_{\leq l}(\Pi_i)|})$ for the number of the shortcut plans k will be explained later as part of the ε-privacy proof.

In summary, all agents sequentially generate solutions to their local problems Π_i at line 5. The systematic local plans are supplemented by longer randomly generated local plans at line 7. Then the agents create public plans by making public projections $\pi^{\triangleright}$ of their generated solutions. Created public plans $\pi^{\triangleright}$ are then stored in a set Φ_i. As the plans π are local solutions, they are α_i-extensible. Agents continuously check whether there are some plans in the intersection of these sets from all other agents. It is important to compute the intersection without disclosing any information about plans which do not belong to this intersection. Plans in the intersection are guaranteed to be α_i-extensible by all agents and thus form global solutions. If at least one global solution is found in Φ, the algorithm ends at line 11. The algorithm ends for all agents in the same iteration, as the secure intersection is done distributively by all agents for all agents. Therefore the termination condition at line 10 is evaluated by all agents equally.

Algorithm 1. ε-Strong privacy preserving multi-agent planner.

1 **Function** SecureMAPlanner(Π_i, ε) **is**
2      $\Phi_i \leftarrow \emptyset$;
3      $l \leftarrow 1$;
4   **loop**
5          $S \leftarrow$ generate all local solutions to Π_i of length l;
           $k \leftarrow |\text{SOLS}_{>l}(\Pi_i)|(1 - \frac{\varepsilon}{|\text{SOLS}_{\leq l}(\Pi_i)|})$; $\varepsilon \leftarrow \varepsilon - k$;
6          $S' \leftarrow$ randomly select k solutions to Π_i of any length $> l$;
7          $\Phi_i \leftarrow \Phi_i \cup \{\pi^{\triangleright} | \pi \in S \cup S'\}$;
8          $\Phi \leftarrow$ secure $\left(\bigcap_{\alpha_j \in \mathcal{A}} \Phi_j \right)$;
9   **if** $\Phi \neq \emptyset$ **then**
10   **return** Φ;
11   **end**
12         $l \leftarrow l + 1$;
13   **end**
14 **end**

The description of the planning algorithm is followed by proofs of its soundness (a result of the algorithm is always a correct solution), completeness (if a planning problem has a solution, it is returned by the algorithm) and assurance on information leakage no more than required ε.

Theorem 1 (Soundness and Completeness). *Algorithm* SecureMAPlanner *is sound and complete, under the assumption that the systematic plan generation procedure (line 5) is complete.*

Proof. **(Soundness)** *Every public plan returned by the algorithm is α_i-extensible by every agent, and thus it can be extended by all agents to a valid global solution (Lemma 1 in [16]).*

    (Completeness) *Since there is only finite number of different plans of length at most l, all plans are eventually (in finite time) added to the plan set Φ_i under the assumption that the underlying plan generation procedure of the local solutions (line 5) is complete. The longest possible solution is finite by Eq. (1), thus* SecureMAPlanner() *with systematic local planner ends in finite time when $\mathcal{M}$ has a solution.*

Theorem 2 (ε-Strong Privacy). *Algorithm* `SecureMAPlanner()` *is ε-strong privacy preserving when ideal PSI is used for the secure plan projection intersection (line 9).*

Proof. The only points in the algorithm, where the agents communicate is in the distributed intersection of the public projections (line 9) and implicitly in the synchronized termination (lines 10–12).

To ensure privacy of the first point, both agents encode public projections of their plans into a set of numbers using the same encoding. Then, they just need to compare two sets of numbers representing their sets of plausible public plans, in other words they need to compute ideal PSI [6,11]. No private information leaks within ideal PSI, therefore no private information leaks during the distributed intersection.

There can be, however, private information leakage, when the algorithm continues several iterations, i.e. the algorithm does not terminate (the second point). When the agent α_- finds out that some set of plans is unacceptable by the agent α_+ (which is the only reason, why the algorithm has to continue with another iteration), private information leaks simply by growth of certainty by Definition 2.

As α_+ does not know how many plans α_- has generated thus how many plans were refused, if we want to upper-bound the possible leaked information, α_+ has to consider that all possible plans of length l were generated by α_- and refused by α_+, i.e. $\lambda_{l+1} - \lambda_l \leq |\text{SOLS}_l(\Pi_i)|$. Such situation reflects the maximal possible growth in the certainty about acceptance of possible plans. In sum over all plans lengths we get $\lambda_l \leq \sum_{1 \leq l' \leq l} |\text{SOLS}_{l'}(\Pi_i)| = |\text{SOLS}_{\leq l}(\Pi_i)|$.

To limit the leakage by shortcutting the solution prematurely by the randomly selected plans, it has to happen that all agents generate by chance a global solution (line 7) sooner than systematically in its iteration by the length l. As the best we can get is an upper-bound (the number of real solution is not known in beforehand) on the needed number of randomly generated plans k, we can assume that there is only one solution plan. A chance to randomly choose one particular plan by a selection of k random plans from all solutions $|\text{SOLS}_{>l}| = n$ longer than the current iteration length l is

$$\frac{\binom{n}{k} - \binom{n-1}{k}}{\binom{n}{k}} = \frac{\binom{n-1}{k-1}}{\binom{n}{k}} = \frac{k}{n} = \frac{k}{|\text{SOLS}_{>l}(\Pi_i)|}. \tag{3}$$

The change of not selecting the solution plan is simply $1 - \frac{k}{|\text{SOLS}_{>l}(\Pi_i)|}$, which with the upper-bound on the certainty about acceptance of possible plans gives us a parameterized tighter upper-bound on the leaked information $\lambda_l \leq |\text{SOLS}_{\leq l}(\Pi_i)|(1 - \frac{k}{|\text{SOLS}_{>l}(\Pi_i)|}) \leq \text{SOLS}_{\leq l}(\Pi_i)$. Note that each agent has the same chance to select the one common solution, therefore the chance is not decreased with increasing number of agents.

By Definition 3, $\lambda \leq \varepsilon$ has to hold for ε-privacy preserving planner, that means for the last iteration with a systematically found global plan $\lambda_{|\pi^*|} \leq \varepsilon$. As the length of the systematic solution plan $|\pi^*|$ is not know in beforehand, we have *to heuristically estimate it. As $l \leq |\pi^*|$ holds for all iterations of the algorithm, we can modify the formula and derive the upper-bound on the number of shortcut plans to fulfill ε:*

$$\lambda_l \leq \varepsilon, \tag{4}$$

$$|\text{SOLS}_{\leq l}(\Pi_i)|(1 - \frac{k}{|\text{SOLS}_{>l}(\Pi_i)|}) \leq \varepsilon, \tag{5}$$

$$|\text{SOLS}_{>l}(\Pi_i)|(1 - \frac{\varepsilon}{|\text{SOLS}_{\leq l}(\Pi_i)|}) \leq k. \tag{6}$$

As such k is only an estimate assuming each iteration is the last one, we have to decrease the allowed leakage in each iteration by $\varepsilon \leftarrow \varepsilon - k$ at line 6.

The parameter ε, and consequentially also k, acts as a trade-off parameter between security and efficiency. If the agent "randomly" selects all its plans $\text{SOLS}_{>l}(\Pi_i)$, then no information about refused plans can leak as it is assured that planning finds the (at least one existing) solution in the first iteration. Thus it would imply the strong privacy, i.e. for $k = |\text{SOLS}_{>l}(\Pi_i)|$ we get $\varepsilon \geq 0 \geq \lambda$ from Eq. 5 and Definition 3. Conversely, if we plan only systematically $k = 0$, the leakage upper-bounded $\varepsilon \geq |\text{SOLS}_{\leq l}(\Pi_i)| \geq \lambda$.

In the previous cases, $\text{SOLS}_{>l}$ can be replaced by $\text{SEQS}_{>l}$, as there cannot be less sequences than solutions and we are dealing with upper bounds. However, we kept the tighter $\text{SOLS}_{>l}$ in the proof and discussion. The possible issue with $|\text{SOLS}_{>l}|$ is that it can be hard to evaluate them efficiently, which is not problem with $\text{SEQS}_{>l}$. The drawback of $\text{SEQS}_{>l}$ is their exponentially larger amount, therefore exponential "looseness" of the bounds and a need for possibly exponentially larger k.

The leakage bounds are illustrated in Fig. 1 for an example planning problem. The problem has 2, 4, 8, 16 and 32 solutions (in the set $\text{SOLS}_{\leq l}(\Pi_i)$) for plan lengths l 1, 2, 3, 4 and 5 respectively. This gives us 30, 28, 24, 16 and 0 solutions in the set $\text{SOLS}_{>l}(\Pi_i)$, again for lengths $l = 1 <$, 2, 3, 4 and 5. The lines depict the upper-bound of the leakage λ for different numbers of shortcut plans k. For example, in the first iteration, only the two possibly refused plans can leak information, therefore even when $k = 0$, i.e. without any shortcut plans, $\lambda_1 = 2$. Conversely, to assure the planning process ends in the first iteration and does not leak any information, k has to equal to the rest of plans for $> l$, which is 30, where maximal leakage is ensured to be 0. Based on the changing numbers of already generated and still to be generated plans the ratio changes with the following iterations.

The points, where the upper-bounds equals to 0, represent numbers of shortcut plans needed to provide strong privacy (Proposition 1). The Fig. 2 depicts the numbers k of shortcut plans for the particular iterations of the example planning problem. Although the example shows only a small and synthetic planning problem, the principles and trends of the privacy bounds are general.

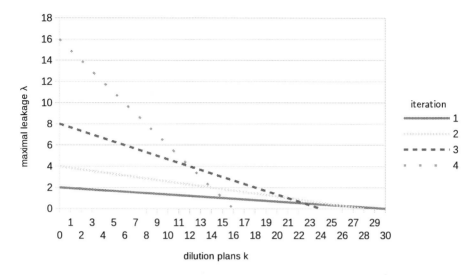

Fig. 1. Upper-bounds of required shortcut plans to assure leakage of the planning algorithm for iterations of the presented example planning problem.

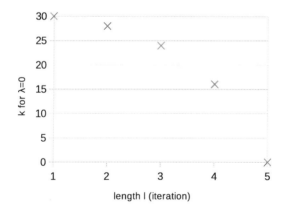

Fig. 2. Amounts of shortcut plans k assuring leakage of no private information, therefore privacy preserving run of the planning algorithm in the presented example planning problem.

To increase efficiency of the intersections, the principle proposed in PSM planner [16] can be used. Each agent stores generated plans in a form of planning state machines, special version of finite state machines. An algorithm, which can be used for secure intersection of planning state machines, was presented in [5]. In the case of different representation of public plans, more general approach of generic secure computation can be applied [1,20,21].

4 Logistics Example

Let us consider a simple logistics scenario to demonstrate how private information can leak for $k = 0$ and how it decreases with larger k values.

In this scenario, there are two transport vehicles (plane and truck) operating in three locations (prague, brno, and ostrava). A plane can travel from prague to brno and back, while a truck provides connection between brno and ostrava. The goal is to transport the crown from prague to ostrava.

This problem can be expressed using MA-STRIPS as follows. Actions

$$\text{fly}(\textit{loc}_1, \textit{loc}_2) \text{ and } \text{drive}(\textit{loc}_1, \textit{loc}_2)$$

describe movement of plane and truck respectively. Actions load($\textit{veh}$, $\textit{loc}$) and unload($\textit{veh}$, $\textit{loc}$) describe loading and unloading of crown by a given vehicle at a given location.

We define two agents $\textit{Plane}$ and $\textit{Truck}$. The agents are defined by sets of executable actions as follows

$$
\begin{aligned}
\textit{Plane} = \{ & \\
& \text{fly(prague, brno), fly(brno, prague),} \\
& \text{load(plane, prague), load(plane, brno),} \\
& \text{unload(plane, prague), unload(plane, brno) } \}, \\
\textit{Truck} = \{ & \\
& \text{drive(brno, ostrava), drive(ostrava, brno),} \\
& \text{load(truck, brno), load(truck, ostrava),} \\
& \text{unload(truck, brno), unload(truck, ostrava) } \}.
\end{aligned}
$$

The aforementioned actions are defined using binary variables at($\textit{veh}$, $\textit{loc}$) $\in$ {true, false} to describe possible vehicle locations and binary variables in(crown, $\textit{loc}$) and in(crown, $\textit{veh}$) to describe positions of crown. A variable is assigned true value if the fact it is describing holds. E.g. in(crown, plane) $=$ true represents the fact that crown is in plane. We omit action names in examples when no confusion can arise. For example, we have the following actions:

$$
\begin{aligned}
\text{fly}(\textit{loc}_1, \textit{loc}_2) = \langle \text{pre}(.) = \{ & \text{at(plane, } \textit{loc}_1) = \text{true}\}, \\
\text{eff}(.) = \{ & \text{at(plane, } \textit{loc}_2) \leftarrow \text{true,} \\
& \text{at(plane, } \textit{loc}_1) \leftarrow \text{false}\}\rangle,
\end{aligned}
$$

$$
\begin{aligned}
\text{load}(\textit{veh}, \textit{loc}) = \quad \langle \text{pre}(.) = \{ & \text{at}(\textit{veh}, \textit{loc}) = \text{true, in(crown, } \textit{loc}) = \text{true}\}, \\
\text{eff}(.) = \{ & \text{in(crown, } \textit{veh}) \leftarrow \text{true,} \\
& \text{in(crown, } \textit{loc}) \leftarrow \text{false}\}\rangle, \\
\text{unload}(\textit{veh}, \textit{loc}) = \langle \text{pre}(.) = \{ & \text{at}(\textit{veh}, \textit{loc}) = \text{true, in(crown, } \textit{veh}) = \text{true}\}, \\
\text{eff}(.) = \{ & \text{in(crown, } \textit{loc}) \leftarrow \text{true,} \\
& \text{in(crown, } \textit{veh}) \leftarrow \text{false}\}\rangle.
\end{aligned}
$$

The initial state and the goal are given as follows:

$$s_I = \{\texttt{at}(\texttt{plane}, \texttt{prague}) = \texttt{true}, \ \texttt{at}(\texttt{truck}, \texttt{brno}) = \texttt{true},$$
$$\texttt{in}(\texttt{crown}, \texttt{prague}) = \texttt{true}\}$$
$$s_\star = \{\texttt{in}(\texttt{crown}, \texttt{ostrava}) = \texttt{true}\}$$

All other variables not present in the initial state s_I are false.

In our example, the only variable shared by the two agents is $\texttt{in}(\texttt{crown}, \texttt{brno})$ and as required by $\texttt{vars}(s_\star) \subseteq \mathcal{V}^{\text{pub}}$ (see Sect. 2), the goal $\texttt{in}(\texttt{crown}, \texttt{ostrava})$. We have the following variable classification:

$$\mathcal{V}^{\text{pub}} = \{\texttt{in}(\texttt{crown}, \texttt{brno}),$$
$$\texttt{in}(\texttt{crown}, \texttt{ostrava})\},$$
$$\mathcal{V}^{\text{priv}}_{Plane} = \{\texttt{at}(\texttt{plane}, \texttt{prague}), \texttt{at}(\texttt{plane}, \texttt{brno}),$$
$$\texttt{in}(\texttt{crown}, \texttt{prague}), \texttt{in}(\texttt{crown}, \texttt{plane})\},$$
$$\mathcal{V}^{\text{priv}}_{Truck} = \{\texttt{at}(\texttt{truck}, \texttt{brno}), \texttt{at}(\texttt{truck}, \texttt{ostrava}),$$
$$\texttt{in}(\texttt{crown}, \texttt{truck})\}.$$

The actions and their projections important for the following discussion are:

$$\texttt{load}(\texttt{truck}, \texttt{brno}) = \langle \text{pre}(.) = \{\texttt{in}(\texttt{crown}, \texttt{brno}) = \texttt{true},$$
$$\texttt{in}(\texttt{truck}, \texttt{brno}) = \texttt{true}\}\rangle,$$
$$\text{eff}(.) = \{\texttt{in}(\texttt{crown}, \texttt{brno}) \leftarrow \texttt{false},$$
$$\texttt{in}(\texttt{crown}, \texttt{truck}) \leftarrow \texttt{true}\},$$
$$\texttt{load}(\texttt{truck}, \texttt{brno})^\triangleright = \langle \text{pre}(.) = \{\texttt{in}(\texttt{crown}, \texttt{brno}) = \texttt{true}\},$$
$$\text{eff}(.) = \{\texttt{in}(\texttt{crown}, \texttt{brno}) \leftarrow \texttt{false}\}\rangle,$$

$$\texttt{unload}(\texttt{truck}, \texttt{ostrava}) = \langle \text{pre}(.) = \{\texttt{in}(\texttt{truck}, \texttt{ostrava}) = \texttt{true},$$
$$\texttt{in}(\texttt{crown}, \texttt{truck}) = \texttt{true}\}\rangle,$$
$$\text{eff}(.) = \{\texttt{in}(\texttt{crown}, \texttt{ostrava}) \leftarrow \texttt{true},$$
$$\texttt{in}(\texttt{crown}, \texttt{truck}) \leftarrow \texttt{false}\},$$
$$\texttt{unload}(\texttt{truck}, \texttt{ostrava})^\triangleright = \langle \text{pre}(.) = \emptyset,$$
$$\text{eff}(.) = \{\texttt{in}(\texttt{crown}, \texttt{ostrava}) \leftarrow \texttt{true} \}\rangle$$

All the actions arranging vehicle movements are private. Only the actions providing package treatment at public locations (brno, ostrava) are public:

$$O^{\text{pub}}_{Truck} = \{ \texttt{load}(\texttt{truck}, \texttt{brno}), \texttt{unload}(\texttt{truck}, \texttt{brno}),$$
$$\texttt{load}(\texttt{truck}, \texttt{ostrava}), \texttt{unload}(\texttt{truck}, \texttt{ostrava}) \},$$

$$O^{\text{pub}}_{Plane} = \{\texttt{load}(\texttt{plane}, \texttt{brno}), \texttt{unload}(\texttt{plane}, \texttt{brno})\}.$$

The agent *Plane* generates possible plans using the systematic plan generation algorithm (e.g. Best-First Search) and thus it sequentially generates following public plans:

$$\pi_1^{Plane} = \langle\ \texttt{unload(truck, ostrava)}\ \rangle, l = 1$$
$$\pi_2^{Plane} = \langle\ \texttt{unload(plane, brno)},$$
$$\texttt{unload(truck, ostrava)}\ \rangle, l = 2,$$
$$\pi_3^{Plane} = \langle\ \texttt{unload(truck, brno)},$$
$$\texttt{unload(truck, ostrava)}\ \rangle, l = 2$$
$$\pi_4^{Plane} = \langle\ \texttt{unload(truck, ostrava)},$$
$$\texttt{unload(truck, ostrava)}\ \rangle, l = 2$$
$$\cdots$$
$$\pi_n^{Plane} = \langle\ \texttt{unload(plane, brno)},\ \texttt{load(truck, brno)},$$
$$\texttt{unload(truck, ostrava)}\ \rangle, l = 3.$$

Note that any locally valid sequence of action containing action

$$\texttt{unload(truck, ostrava)}$$

seems to be a valid solution to the *Plane* agent. In this example, π_n^{Plane} is the first plan extensible to a global solution by both *Plane* and *Truck* generated by the systematic planning process.

Similarly, agent *Truck* sequentially generates following public plans:

$$\pi_1^{Truck} = \langle\ \texttt{unload(plane, brno)},\ \texttt{load(truck, brno)},$$
$$\texttt{unload(truck, ostrava)}\ \rangle, l = 3,$$
$$\pi_2^{Truck} = \langle\ \texttt{unload(plane, brno)},\ \texttt{unload(plane, brno)},$$
$$\texttt{load(truck, brno)},\ \texttt{unload(truck, ostrava)}\ \rangle, l = 4,$$
$$\cdots$$

We can see that *Truck* generates an extensible plan as the first one and *Plane* generated equivalent solution as the n-th plan. Thus, once both agents agree on a solution, agent *Plane* can try to deduce something about *Truck* private information. Since all plans $\pi_1^{plane}, \ldots, \pi_4^{plane}$ are strictly shorter than the accepted solution π_n^{plane} and they were not generated by *Truck*, it implies that these plans are not acceptable by *Truck*, i. e. for example $P^*(\pi_1^{Plane}) = 0$. More specifically, *Plane* can deduce following about *Truck*'s private information:

- The action $\texttt{unload(truck, ostrava)}$ has to contain some private precondition, otherwise π_1^{Plane} would be generated also by *Truck* before π_1^{Truck}, because it is shorter.
- Private preconditions of $\texttt{unload(truck, ostrava)}$ certainly depend on private fact (possibly indirectly) generated by $\texttt{load(truck, brno)}$, otherwise π_2^{Plane} would be generated before π_1^{Truck}.

In this example, we have shown how systematic generation of plans can cause private information leakage. Let us now consider a case when both agents add one shortcut plan after each systematically generated one, i. e. $k = 1$. For the simplicity, we will consider previous sequence of plans, where π_n^{Plane} is selected as the shortcut plan in the first iteration for $l = 1$ by both agents. In such case, the amount of leaked information is smaller by Eq. (5). If there is only one solution π_n^{Plane}, the leakage will be 0 and the agents would not be able to deduce any private information about the other agents. If the shortcut plan π_n^{Plane} is added in next iteration for length $l = 3$, *Plane* can deduce that $P^*(\pi_1^{Plane}) = 0$, but cannot deduce the same about other plans. *Plane* could deduce that *Truck* accepts no plan of length 2, only once it is sure that all of them have been systematically generated. But thanks to the adding of the shortcut plan, the solution can be found sooner.

Obviously $k = 1$ decreases the leaked information only minimally. To decrease the private information leakage significantly, k has to grow by Eq. (5) towards $|\text{SOLS}_{>l}(\Pi_{Plane})|$ as we showed in proof of Theorem 2.

5   Code Lock Example and General Privacy Leakage Upper-Bound

The other example is designed such that it shows the successive leaking of information by learning about refused sequences of the actions. There is a combination code lock and two agents. The agent α_- is unlocking the lock with help of the other agent α_+, which knows the combination. The lock requires a correct sequence of n pressed buttons reachable by α_+ (each button can be pressed only once), finished with pressing two unlock buttons, each reachable only by one of the agents. Note that strictly speaking if only one combination is correct no private information would leak as Definition 2 excludes the solution plan π^*. We could modify the example such that there are more correct solution and α_- is trying to deduce all of them, however to simplify the latter discussion, we will stick to one solution, which we assume to be secret.

This assumption is not unrealistic, as in reality the press actions would be private, however by the requirement of privacy preserving MA-MPT planning, all actions incl. press are public.

The binary variables of the problem are

$$\mathcal{V}^{\text{pub}} = \{\texttt{unlocked}_+, \texttt{unlocked}_-\},$$
$$\mathcal{V}_{\alpha_+}^{\text{priv}} = \{\texttt{pressed}_0, \texttt{pressed}_1, \ldots, \texttt{pressed}_n\},$$
$$\mathcal{V}_{\alpha_-}^{\text{priv}} = \emptyset.$$

The two public unlocked variables describe whether the two agents successfully unlocked their side of the lock. For simplicity, we assume only α_+ need to enter the correct sequence, which allows to unlock its side. The agent α_- attempts to deduce the constraints among the presses during the process. Provided that the α_- agent has similar combination as α_+ the example would work symmetrically for both agents, or even for more agents unlocking the lock in coordination. The successfully entered steps of the combination are represented by the private pressed variables.

In the initial state, all variables are set to false with the exception of $\texttt{pressed}_0$ allowing to press the first correct button. The goal of the problem is to unlock both sides of the lock:

$$s_I = \{\texttt{pressed}_0 = \texttt{true}\},$$
$$s_\star = \{\texttt{unlocked}_+ = \texttt{true}, \texttt{unlocked}_- = \texttt{true}\}.$$

The actions of the problem are the two unlocking the lock and actions representing pressing the buttons. All actions are public (following the assumption on no private actions), however actions of α_+ have private preconditions and effects constraining only the correct unlock sequence. The action sets consist of:

$$O_{\alpha_+}^{\text{pub}} = \{\text{unlock}_+, \text{press}_1, \text{press}_2, \ldots, \text{press}_n\},$$
$$O_{\alpha_-}^{\text{pub}} = \{\text{unlock}_-\}.$$

The action unlock_- (and its projection) has no preconditions and the sole effect $\text{unlocked}_- \leftarrow \text{true}$, which is required by the goal:

$$\text{unlock}_- = \text{unlock}_-^{\triangleright} = \langle \text{pre}(.) = \emptyset, \text{eff}(.) = \{\text{unlocked}_+ \leftarrow \text{true}\}\rangle.$$

Let the unlocking sequence of button indices be described by a mapping $u(i) \mapsto i'$ which for each step i returns next button index i' to be pressed. Then each button pressing actions is defined as:

$$\text{press}_i = \langle \text{pre}(.) = \{\text{pressed}_{u(i)-1} = \text{true}\},$$
$$\text{eff}(.) = \{\text{pressed}_{u(i)} \leftarrow \text{true}\}\rangle.$$
$$\text{unlock}_+ = \langle \text{pre}(.) = \{\text{pressed}_n = \text{true}\},$$
$$\text{eff}(.) = \{\text{unlocked}_+ \leftarrow \text{true}\}\rangle.$$

For an example, a $n = 3$ step unlocking sequence $2, 3, 1$ will induce following actions for the agent α_+:

$$\text{press}_2 = \langle \text{pre}(.) = \{\text{pressed}_0 = \text{true}\},$$
$$\text{eff}(.) = \{\text{pressed}_1 \leftarrow \text{true}\}\rangle,$$
$$\text{press}_3 = \langle \text{pre}(.) = \{\text{pressed}_1 = \text{true}\},$$
$$\text{eff}(.) = \{\text{pressed}_2 \leftarrow \text{true}\}\rangle,$$
$$\text{press}_1 = \langle \text{pre}(.) = \{\text{pressed}_2 = \text{true}\},$$
$$\text{eff}(.) = \{\text{pressed}_3 \leftarrow \text{true}\}\rangle.$$
$$\text{unlock}_+ = \langle \text{pre}(.) = \{\text{pressed}_3 = \text{true}\},$$
$$\text{eff}(.) = \{\text{unlocked}_+ \leftarrow \text{true}\}\rangle.$$

The projections of all press actions have no preconditions and effects (they are effectively noops with different action names from perspective of α_-), as the pressed variables are private. The action unlock has only its goal effect:

$$\text{press}_1^{\triangleright}, \ldots, \text{press}_n^{\triangleright} = \langle \text{pre}(.) = \emptyset, \text{eff}(.) = \emptyset\rangle = \text{noop},$$
$$\text{unlock}_+^{\triangleright} = \langle \text{pre}(.) = \emptyset, \text{eff}(.) = \{\text{unlocked}_+ \leftarrow \text{true}\}\rangle.$$

From perspective of α_-, any sequence of α_+ ending with unlock_+ is legitimate. On the contrary, only the correct sequence of press actions and unlock_+ is valid from perspective of α_+.

In each iteration l, if a solution is not found, α_- eliminates all possible combinations of the code of length l. These eliminations represent the private variables pressed in form of preconditions and effects of actions of α_+. Because of the

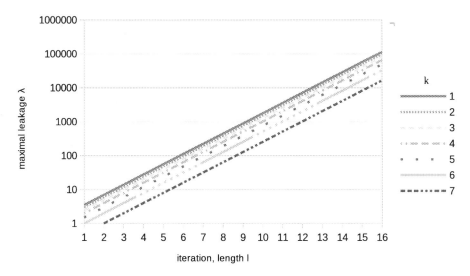

Fig. 3. Hypothetical upper-bounds on information leakage in the Code Lock example problem for $n = 2$ without decreasing ε during the iterations by $\varepsilon \leftarrow \varepsilon - k$. For $k = 8$ the leakage $\lambda = 0$. For $l \geq 3$, the leakage assumes possible longer plans of the problem (e.g., by more then two agents) than those limited by $l \leq n$.

problem formulation, the number of projected local plans of α_+ from perspective of α_- are:

$$n^1 + n^2 + \cdots + n^l + n^{l+1} + \cdots + n^{n-1} + n^n,$$

where n is the number of **press** actions and l is the intermediate length of the solution. We can bound sizes of $\text{SOLS}_{\leq l}$ and $\text{SOLS}_{>l}$ using this sequence, $l \leq n$ by the count of meaningful plans by Eq. 1 and an assumption on repetitions of the **press** actions are allowed as α_- cannot know otherwise:

$$|\text{SOLS}_{\leq l}| \leq \sum_{i=1}^{l} n^i \leq n^{l+1}, \tag{7}$$

$$|\text{SOLS}_{>l}| \leq \sum_{i=1}^{n-l} n^{l+i} \leq n^{n+1}. \tag{8}$$

Without the shortcut plans $k = 0$, we get that $n^{l+1} \leq \varepsilon$ using Eqs. (5) and (7). This shows the possible maximal leakage of information is exponentially bound by the length of the plan which is bound by the number of actions in Π_+. It is not surprising though that if we want a strong privacy preserving variant $\varepsilon = 0$, we need to generate exponential number of plans n^{n+1} in the number of actions of the problem for the first iteration by Eqs. (6) and (8).

The exponential growth of information leakage in l in this example is illustrated in Fig. 3. The other exponential dependency is on the number of (**press**) actions, i.e. on n.

As the Code Lock problem is designed such that there are no public dependencies among the actions (with exception of the required goal conditions), it represents a planning problem with the maximal amount of private information and only one solution as assumed in the proof of Theorem 2. Therefore the resulting exponential bounds on leakage in l and in n hold not only for this particular planning problem, but as a general upper bound on the privacy leakage in MA-STRIPS planning with n actions and k shortcut plans in l-th iteration of the `SecureMAPlanner()` algorithm (by Eqs. (5), (7) and (8)):

$$\lambda_l \leq n^{l+1}(1 - \frac{k}{n^{n+1}}).\tag{9}$$

6 Conclusions

In this article, we have provided a refined variant of the multi-agent planning principle preserving privacy from our previous version of the paper [17]. The principle of the algorithm is an application of the private set intersection (PSI) algorithm to privacy preserving multi-agent planning using intersection of sets of plans. As the plans are generated as extensible to a global solution provided that all agents agree on a selection of such local plans, the soundness of the planning approach is ensured. As we showed in [17] and in the refined proof of Theorem 2 here, the intersection process can be secure in one iteration by PSI, but some private information can leak during iterative generation of the local plans, which is the only practical way how to solve generally intractable planning problems. In more iterations, plans which are extensible by some agents but not extensible by all agents can leak private information about private dependencies of actions within the plans. In other words, if an agent says the proposed solution can be from its perspective used as a solution to the planning problem, but it cannot be used as a solution by another agent, the first one learns that the other one needs to use some private actions which obviate usage (extensibility) of the plan to a global solution. In the previous version of the algorithm, we have proposed to dilute the plans by an sufficient amount of randomized plans, however the number in general needed to be infinite [17]. In this variant of the algorithm, we have shown that the privacy leakage can be arbitrarily shortcut by randomly selected local plans and fully prevented by using all solutions (exponential in the number of actions) already in the first iteration. Although the number of such shortcut plans achieving strong privacy is exponential in general, in contrast to the dilution approach, the number is finite. The results are also in agreement with our recent results in [14]. As in the previous version of the paper, we have illustrated the principle on the logistics example, however with new results using the improved version of the algorithm. Newly, we have demonstrated the principle on a synthetic planning problem, which shows also the novel privacy leakage upper bound (Eq. 9) which holds in general.

Acknowledgments. This research was supported by the Czech Science Foundation (no. 15-20433Y).

References

1. Ben-Or, M., Goldwasser, S., Wigderson, A.: Completeness theorems for non-cryptographic fault-tolerant distributed computation. In: Proceedings of the Twentieth Annual ACM Symposium on Theory of Computing, STOC 1988, pp. 1–10. ACM, New York (1988)
2. Brafman, R.I.: A privacy preserving algorithm for multi-agent planning and search. In: Yang, Q., Wooldridge, M. (eds.) Proceedings of the Twenty-Fourth International Joint Conference on Artificial Intelligence, IJCAI 2015, Buenos Aires, Argentina, 25–31 July 2015, pp. 1530–1536. AAAI Press (2015)
3. Brafman, R.I., Domshlak, C.: From one to many: planning for loosely coupled multi-agent systems. In: Proceedings of the ICAPS 2008, pp. 28–35 (2008)
4. Diffie, W., Hellman, M.: New directions in cryptography. IEEE Trans. Inf. Theor. **22**(6), 644–654 (1976)
5. Guanciale, R., Gurov, D., Laud, P.: Private intersection of regular languages. In: Miri, A., Hengartner, U., Huang, N., Jøsang, A., García-Alfaro, J. (eds.) 2014 Twelfth Annual International Conference on Privacy, Security and Trust, Toronto, ON, Canada, 23–24 July 2014, pp. 112–120. IEEE (2014). https://doi.org/10.1109/PST.2014.6890930
6. Jarecki, S., Liu, X.: Fast secure computation of set intersection. In: Garay, J.A., De Prisco, R. (eds.) SCN 2010. LNCS, vol. 6280, pp. 418–435. Springer, Heidelberg (2010). https://doi.org/10.1007/978-3-642-15317-4_26
7. Maliah, S., Shani, G., Brafman, R.: Online macro generation for privacy preserving planning. In: Proceedings of the 26th International Conference on Automated Planning and Scheduling, ICAPS 2016 (2016)
8. Maliah, S., Shani, G., Stern, R.: Collaborative privacy preserving multi-agent planning. In: Proceedings of the AAMAS 2016, pp. 1–38 (2016)
9. Nissim, R., Brafman, R.I.: Multi-agent A* for parallel and distributed systems. In: Proceedings of AAMAS 2012, pp. 1265–1266 (2012)
10. Nissim, R., Brafman, R.I.: Distributed heuristic forward search for multi-agent planning. JAIR **51**, 293–332 (2014)
11. Pinkas, B., Schneider, T., Segev, G., Zohner, M.: Phasing: Private set intersection using permutation-based hashing. In: 24th USENIX Security Symposium (USENIX Security 15), pp. 515–530. USENIX Association, Washington, D.C. (2015)
12. Štolba, M., Tožička, J., Komenda, A.: Secure multi-agent planning. In: Proceedings of the International Workshop on PrAISe (2016)
13. Stolba, M., Tozicka, J., Komenda, A.: Secure multi-agent planning algorithms. ECAI **2016**, 1714–1715 (2016)
14. Stolba, M., Tozicka, J., Komenda, A.: The limits of strong privacy preserving multi-agent planning. In: Proceedings of the 27th International Conference on Automated Planning and Scheduling. ICAPS 2017 (2017). To appear
15. Torreño, A., Onaindia, E., Sapena, O.: FMAP: distributed cooperative multi-agent planning. AI **41**(2), 606–626 (2014)
16. Tožička, J., Jakubův, J., Komenda, A., Pěchouček, M.: Privacy-concerned multi-agent planning. KAIS, pp. 1–38 (2015)
17. Tozicka, J., Komenda, A., Stolba, M.: ε-strong privacy preserving multiagent planner by computational tractability. In: van den Herik, H.J., Rocha, A.P., Filipe, J. (eds.) Proceedings of the 9th International Conference on Agents and Artificial Intelligence. ICAART 2017, vol. 1, Porto, Portugal, 24–26 February 2017, pp. 51–57. SciTePress (2017). https://doi.org/10.5220/0006176400510057

18. Van Der Krogt, R.: Quantifying privacy in multiagent planning. Multiagent Grid Syst. **5**(4), 451–469 (2009)
19. Štolba, M., Komenda, A.: Relaxation heuristics for multiagent planning. In: 24th International Conference on Automated Planning and Scheduling (ICAPS), pp. 298–306 (2014)
20. Yao, A.C.: Protocols for secure computations. In: Proceedings of the 23rd Annual Symposium on Foundations of Computer Science, pp. 160–164. SFCS 1982. IEEE Computer Society, Washington, DC (1982)
21. Yao, A.C.C.: How to generate and exchange secrets. In: Proceedings of the 27th Annual Symposium on Foundations of Computer Science, SFCS 1986, pp. 162–167. IEEE Computer Society, Washington, DC (1986)

A Quantitative Assessment of the JADEL Programming Language

Federico Bergenti[1], Eleonora Iotti[2(✉)], Stefania Monica[1], and Agostino Poggi[2]

[1] Dipartimento di Scienze Matematiche, Fisiche e Informatiche,
Università degli Studi di Parma, 43124 Parma, Italy
{federico.bergenti,stefania.monica}@unipr.it
[2] Dipartimento di Ingegneria e Architettura,
Università degli Studi di Parma, 43124 Parma, Italy
eleonora.iotti@studenti.unipr.it, agostino.poggi@unipr.it

Abstract. This paper reports a quantitative assessment of JADEL, an agent-oriented programming language designed to implement JADE agents and multi-agent systems. The assessment is structured in two parts. The first part is intended to evaluate the effectiveness of JADEL for the concrete implementation of agent-based algorithms expressed using a pseudocode. The second part examines the functionality of the language regarding concurrency and message passing by comparing the implementation in JADEL of a set of benchmark algorithms with the corresponding implementations in Scala. The metrics introduced for the two parts of the assessment are meant to evaluate the expressiveness and ease of use of JADEL, and reported results are encouraging.

Keywords: Agent Oriented Programming · JADEL · JADE

1 Introduction

Agent-Oriented Programming (*AOP*) is the programming paradigm first introduced by Shoham in [40]. AOP identifies as core abstractions the autonomous and proactive entities known as (*software*) *agents*, and, over the years, several languages and tools have been developed to coherently support AOP and to provide advanced features for the development of agents and multi-agent systems. *Agent programming languages* is a class of programming languages, which includes AOP languages, that has gained significant relevance in the literature. The interest in agent programming languages dates back to the introduction of agent technologies and, since then, it has grown rapidly. As a matter of fact, agent programming languages turned out to be especially convenient to model and develop complex multi-agent systems, in contrast with traditional (lower-level) languages, that are often considered not suitable to effectively implement *Agent-Oriented Software Engineering* (*AOSE*) [13]. Nowadays, agent programming languages are widely recognized as important tools in the development of agent technologies and they represent an important topic of research

© Springer International Publishing AG, part of Springer Nature 2018
J. van den Herik et al. (Eds.): ICAART 2017, LNAI 10839, pp. 157–178, 2018.
https://doi.org/10.1007/978-3-319-93581-2_9

(see, e.g., [1,18]). Each agent programming language is usually based on a specific agent model, which is often formally defined, and aims at providing dedicated constructs to adopt the specific agent model at a high level of abstraction. Simplicity and ease of use are characteristics which made the success of agent programming languages among developers. In fact, thanks to such characteristics, agent programming languages allow developers to reduce the complexity of their work and to expedite the creation of agents and multi-agent systems.

Besides agent programming languages, other tools have been provided over the years to support the effective construction of agents and multi-agent systems. Agent platforms are examples of such tools which offer language-agnostic approaches to the development of agents and multi-agent systems. One of the most popular agent platforms is the *Java Agent DEvelopment framework* (*JADE*, jade.tilab.com) [2,3], which is a middleware that offers several APIs and graphical tools to support the development of distributed multi-agent systems. JADE can be considered a consolidated tool, and it is widely used both for industrial and academic purposes [30]. In particular, it has been used for many relevant research projects (see, e.g., [8,9,11,35], just to mention some recent projects of the authors), and it has been in daily use for service provision and management in Telecom Italia for more than six years, serving millions of customers in one of the largest broadband networks in Europe [10]. Moreover, JADE is considered a valid enabler for the use of agent technology in various application domains, such as agent-based social networks modeling [12] and localization [6,32,33]. As a notable feature, JADE supports the development of agents and multi-agent systems that are compliant with the specifications of *IEEE Foundation for Intelligent and Physical Agents* (*FIPA*, www.fipa.org), with a particular focus on *FIPA interaction protocols* [25].

Beside such considerations, JADE also owes its success to its pure Java approach to agent technologies. As a matter of fact, when JADE was conceived and developed, in the early 2000's, its main design decisions were based on the technologies that were most popular and promising at the time. Developers wanted to use Java, and the common opinion was that such a technology would have been able to change many aspects of software development processes. In such a context, a pure Java approach seemed to be a perfect choice for a software framework that aimed at becoming a solid and reliable tool. Nowadays, such a design choice is less appealing to developers of agents and multi-agent systems. Our experience in using agent technologies and teaching it to graduate students shows a slow regression of the appreciation on JADE for its intimate link with Java. In fact, Java does not natively support agent-oriented technologies and methodologies, and this is perceived as a limitation and a source of errors. Moreover, JADE is constantly expanding and its continuous growth—in terms of features, and available APIs—increases the complexity of the framework. As a result, there is a high number of implementation details that a developer must handle to build a non-trivial multi-agent system. In order to address such problems, we are working on automated tools to help the analysis and verification of JADE agents and multi-agent systems on the basis of a formal operational semantics that we are developing [5, 16].

In a plan of simplification and renovation of the experience of using JADE, the introduction of a specific AOP language seems an appropriate option. The *JADE Language* (*JADEL*) is an AOP language based on JADE which is meant to simplify the work of JADE users, at least, in some aspects of the development process. JADEL provides abstractions and constructs which focus on relevant agent-oriented features of JADE, and it aims at simplifying the adoption of such features at a high level of abstraction. A first description of JADEL can be found in [7], and more recent developments are discussed in [17], where an overview of current syntax and semantics is presented. Then, in [4, 14], the JADEL support for FIPA interaction protocol, which was not included in the first versions of the language, is presented. Finally, [15] is a first attempt at assessing the features of the language by discussing the use of JADEL for the implementation of a non-trivial agent-based algorithm. The objective of this last exercise is to illustrate the steps that a programmer needs to follow from pseudocode to implementation, and to analyze the effort spent in doing such a task when JADEL is adopted. Due to the distributed nature of JADE, the algorithm chosen as case study is a well-known procedure for solving distributed constraint satisfaction problems, the *Asynchronous BackTracking* (*ABT*) algorithm. In [15], the source code written with JADEL of the ABT algorithm is compared with the original pseudocode from [44, 45], and some considerations on the effectiveness of the use of JADEL for this task are presented. In this paper, the problem of translating a known pseudocode into a working program, and of evaluating the effectiveness of JADEL in such a task, is further investigated. The main steps of the translation from the original pseudocode to JADEL source code are recalled, and relevant metrics for the evaluation of programming languages are applied to this case. Moreover, in order to investigate the features of JADEL in terms of support to concurrency and message passing, accepted benchmark algorithms found in the Savina benchmark suite [29] are implemented in JADEL and compared against known implementations in Scala.

The paper is organized as follows. Section 2 briefly reports on some of the most popular agent programming languages to overview related work. Section 3 shows the JADEL programs used to support the proposed quantitative evaluation. In particular, it shows the implementation of the ABT pseudocode and of selected programs of the Savina benchmark suite. Section 4 uses the presented JADEL source codes to discuss a quantitative evaluation of JADEL. Finally, a brief recapitulation of major presented results concludes the paper.

2   Related Work

The obvious collocation of JADEL is in the wide scope of agent programming languages, but JADEL is also a *Domain-Specific Language* (*DSL*) and it should be treated as such. The wide range of technologies and tools involved in the development of a DSL is brightly discussed in [31, 34], where DSLs are clearly marked as important tools to support *model-driven development*. Such works also provide in-depth analyses of the motivations that may lead developers to decide in favor of a new DSL, which is a difficult decision because of the inherent costs of DSLs in terms of implementation and maintenance. Nevertheless, the use of a DSL for a specific application domain leads to important benefits. As a matter of fact, the syntax of a DSL is tailored on the specific domain that it describes, with the aid of user-friendly notations that are simpler than the respective general-purpose ones. This facilitates code understanding, and it allows many repetitive and tedious activities to be automated. Moreover, DSLs are meant to be easily integrated with a host language, which is typically a general-purpose programming language, and this fact ensures the applicability and reusability of domain-specific code in real-world scenarios, where the interoperability with existing code is essential. All such benefits justify the design of JADEL as a DSL for AOP with the intent to increase the adoption of JADE in model-driven development. Notably, the approach of designing an new agent programming language as a DSL for agent-oriented programming has been adopted by other languages mentioned below.

The features of agent programming languages may differ significantly, concerning, e.g., the selected agent mental attitudes (if any), the integration with an agent platform (if any), the underlying programming paradigm, and the underlying implementation language. In order to compare the characteristics of different agent programming languages and to provide a clear overview of the state of the art, it is worth recalling accepted classifications of relevant agent programming languages that have already been proposed. [1] classifies agent programming languages on the basis of the use of mental attitudes. According to such a classification, agent programming languages can be divided into: AOP languages, *Belief Desire Intentions* (*BDI*) languages, hybrid languages, which combine the two previous classes, and other languages, which fall outside previous classes. It is worth noting that such a classification recognizes that BDI languages follow the AOP paradigm, but it reserves special attention to them for their notable relevance in the literature. [18] proposes a different classification, where languages are divided into declarative, imperative, and hybrid. Declarative languages are the most common because they focus on automatic reasoning, which is theme closely related to agent technologies. Some relevant imperative languages have also been proposed, and most of them were obtained by adding agent-oriented constructs to existing procedural programming languages. In the rest of this section, a list of the most popular agent programming languages and their features is given, in chronological order.

Shoham introduced the AOP paradigm in [39] together with his appreciated AGENT-0 language [38]. One of the direct descendant of AGENT-0 is the language called *PLAnning Communicating Agents* (*PLACA*). It extends the capabilities of AGENT-0 by providing improved syntax and new mental categories. Just like AGENT-0, PLACA has experimental nature and it was not meant for practical use. Another important, yet experimental, language is Concurrent METATEM [24], which is an agent programming language based on temporal logics. Another important example of classic agent programming languages is AgentSpeak(L), whose syntax and semantics were formalized by Rao in [36]. The proposed formalization of AgentSpeak(L) is based on the BDI agent model. Other agent programming languages based on the BDI model are *An Abstract Agent Programming Language (3APL)* [27], which includes features of both imperative and logic programming languages, and the *JACK Agent Language (JAL)*, which is built on top of JACK platform [41], an environment to develop multi-agent systems in which agents are based on the BDI paradigm. Another software framework that implements a BDI-based reasoning engine is Jadex [19]. It combines declarative and imperative approaches by using an XML specification language to define beliefs, goals and plans, and by using Java as procedural language to implement plans. Then, *A Computational Language for Autonomous Intelligent and Mobile Agents (CLAIM)* [23] is an agent language that supports agent mobility. While, the *Semantic web-Enabled Agent Language (SEA_L)* [20–22] is a DSL to model and develop multi-agent systems in the scope of the Semantic Web. Finally, SARL [37] is one of the latest entries in the plethora of agent programming languages. It is a general-purpose imperative language with an intuitive syntax, and it can be considered platform-agnostic, even if it is commonly used with the dedicated agent platform called Janus.

3   Implementations of Selected Algorithms in JADEL

JADEL supports four main agent-oriented abstractions, namely, *agents*, *behaviours*, *communication ontologies*, and *roles* in interaction protocols. Actually, agents in JADEL use ontologies and behaviours, and they take roles in interaction protocols.

A JADEL agent can be defined by using the keyword `agent` followed by its name. It has a life cycle that consists in a start-up phase followed by an execution phase, and it is eventually terminated by a take-down phase. The declaration of an agent is allowed to extend the declaration of another agent, with the usual semantics of inheritance, and two event handlers are provided to support initialization and take-down phases, namely, the `on-create` and the `on-destroy` handlers. During agent initialization, a sequence of tasks, called behaviours following accepted JADE nomenclature, can be added to an internal list by means of the `activate-behaviour` expression. After the start-up phase, actions specified in such behaviours are performed by the agent. New tasks can be added dynamically during the life cycle of the agent, and tasks that are no longer needed can be removed.

Behaviours for JADEL can be of two types, namely `cyclic` or `oneshot`. Cyclic behaviours represent actions that remain in the behaviour list of an agent after their execution. This means that the action of a cyclic behaviour can be used one or more times during the life cycle of the agent. A one-shot behaviour, instead, contains an action which terminates immediately and it is removed from the list of the agent after just one execution. The action of a behaviour can be an auto-triggered action, i.e., it starts immediately after its behaviour is chosen by the agent, or it can be triggered by an event, e.g., the reception of a message. Message reception is handled by means of a specific construct of JADEL, the on-when-do construct, which also provides a control over the type of message the agent intends to receive.

Ontologies and interaction protocols are used in agent communication. In particular, ontologies provide formal means to support the semantics of the adopted agent communication language. Ontologies represent one of the most tedious and error prone tasks in the development of multi-agent systems with JADE, and JADE users tend to agree that the large amount of implementation details and repetitive idioms involved in ontology classes shift the focus on technical parts rather than on the semantics of involved ontology elements. For this reason, JADEL provides a special lightweight syntax for ontologies and it permits the automation of many repetitive tasks. As a matter of fact, a JADEL ontology is defined as a set of concepts, predicates, propositions and actions. Such terms compose a sort of dictionary, which is usually organized in a hierarchical structure. Agents sharing such a dictionary can interact by using common terms as content of their messages.

Besides ontologies, JADEL support structured communication by means of a specific constructs to allow agents taking roles in FIPA interaction protocols. Roles are particular behaviours, which are composed of a set of predefined event handlers. Each of such handlers covers a different step of the interaction protocol, by filtering messages through their performatives, as expected from FIPA specifications.

3.1 Implementation of the ABT Algorithm

The *Asynchronous BackTracking (ABT)* [43] algorithm is a well known algorithm to solve *Distributed Constraint Satisfaction Problems (DCSPs)* [43]. DCSPs are distributed variants of constraint satisfaction problems and, as such, a DCSP consists in a finite set of variables and a finite set of constraints over such variables. As in [43], we denote variables as $x_1, x_2, \ldots, x_n$. Each variable x_i takes values in a domain, called D_i. Constraints subsets of $D_1 \times \cdots \times D_n$, and a DCSP is *solved* if and only if a value is assigned to each variable, and each assignment satisfies all constraints. In a DCSP constraints and variables are distributed among agents. Such agents manage a number of variables and they know the constraints over managed variables. Commonly, each agent is associated with just one variable, and it finds an *assignment* of its variable, i.e., a pair (x_i, d) where $d \in D_i$, that satisfies involved constraints. The interactions in the multi-agent system allows each agent to obtain the assignments of other agents, and to check if constraints are really satisfied. Informally, a DCSP is solved if each agent finds a local solution that is consistent with the local solutions of other agents. In [44], a survey of the main algorithms for solving DCSPs is given. In particular, pseudocode and examples are shown for the ABT, the asynchronous weak-commitment search, the distributed breakout, and the distributed consistency algorithms.

The ABT algorithm solves DCSPs under three assumptions: each agent owns exactly one variable, all constraints are in the form of binary predicates, and each agent knows only the constraints that involve its variable. Because it is not necessarily true that all agents in a multi-agent system know each others, they can communicate only if there is a connection between the sender and the receiver of a message. For each agent, the agents who are directly connected with it are called its *neighbors*. In ABT, each agent maintains an *agent view*, which is the agent local view of its neighbors assignments. Communication is addressed by using two types of messages, *OK* and *NoGood*, which work as tools to exchange knowledge on assignments and constraints. More precisely, OK messages are used to communicate the current value of the sender agent variable, and NoGood messages provide the recipient with a new constraint. Agents are associated with a priority order, which can be, e.g., the alphabetical order of their names (or variables). OK messages flow from top to bottom of the priority list of agents, and NoGood messages, instead, go up from lowest priority agents to highest ones. Core of the algorithm is the *check agent view* procedure, which controls if the current known assignments are consistent with the agent value. If not, procedure *backtrack* is used to send NoGood constraints to neighbors. The rest of the algorithm is given in terms of event handling constructs which react at other agents messages.

The ABT algorithm was originally described using a pseudocode [44]. For the sake of brevity, the pseudocode is not reproduced here. The proposed implementation in JADEL follows precisely the original pseudocode. The presentation of the JADEL source code is structured into the presentation of the ontology, of the agents, of support procedures and of event handlers, as follows.

Ontology. An important entity in JADEL is the ontology. From ABT pseudocode, messages are divided into different categories, but there is no specification or definition of an ontology. JADEL takes advantages from a light syntax for defining communication means which describes how agents could interoperate in a given application. The ontology for ABT algorithm includes propositions, concepts, and predicates, as shown in the JADEL code below.

```
ontology ABTOntology {
    concept Assignment(aid index, integer value)
    predicate OK(Assignment assignment)
    predicate NoGood(many Assignment assignmentList)
    proposition NoSolution
    proposition Neighbor
    predicate Solution(many Assignment assignmentList)
}
```

The assignment is a central concept in ABT algorithm. Its implementation consists in the definition of an ontology term which is composed of an agent identifier, i.e., x_i, and the value of its variable, i.e. d_i, called `index` and `value`, respectively. The two predicates used in the main part of the algorithm, namely, the `OK` and the `NoGood` predicates, are defined on the basis of the definition of the

`Assigment`. In fact, an OK message is the current assignment of the agent, while the NoGood message is a sequence of forbidden assignments. Also a predicate `Solution` is defined, which is used to communicate to other agents the solution of the problem, when found. `NoSolution` and `Neighbor` are simply propositions, that agents can exchange to indicate the algorithm termination with no solutions, and the neighbor request, respectively.

Agents. ABT pseudocode describes event handlers and main procedures, but it does not illustrate how agents should be written. In JADEL, an agent must be defined. Such an agent is called `ABTAgent`. It consists of some properties, among which there are the agent view and the set of neighbors. The initialization of an `ABTAgent` is done by filling the set of neighbors with the identifiers of connected agents, and by setting the priority of the each agent. Moreover, `ABTAgent` provides two important methods, namely, `checkConstraints` and `assignVariable`. The first checks if all constraints are satisfied by current assignments in agent view, while the second selects a value which is consistent with agent view and assigns it to the variable owned by the agent. Both methods return `true` if the operation was successful and `false` if it is was not.

Procedures. The core procedure of the ABT algorithm is the *check agent view* procedure, which controls if the current value $my\_value \in D_i$ of the agent x_i is *consistent* with its agent view. A value $d \in D_i$ is called consistent with the agent view if for each value in agent view, all constraints that involve such value and d are satisfied. If this is not the case, the agent has to search for another value. At the end, if none of the values in D_i satisfies the constraints, another procedure is called, namely, the *backtrack* procedure. Otherwise, an OK message is sent to the agent neighbors, which contains the new assignment. In the JADEL implementation of the ABT algorithm, the *check agent view* procedure becomes a one-shot behaviour. In fact, its action has to be performed only once, when the behaviour activates, as follows.

---

oneshot behaviour CheckAgentView **for** ABTAgent {

The keyword `for` denotes which agents are allowed to activate such a behaviour. In this case, such agents are instances of the `ABTAgent` class. Inside the behaviour, methods and public fields of the agent can be called by using the field `theAgent`, which is implicitly initialized with an instance of the agent specified. If no agent is specified with the `for` keyword, `theAgent` refers to a generic agent. The `CheckAgentView` behaviour does not need to wait for messages, or events, so the keyword `do` is used, as follows.

```
do {
    if (!theAgent.checkConstraints()) {
        if (!theAgent.assignVariable()) {
            activate behaviour Backtrack(theAgent)
        } else {
            activate behaviour SendOK(theAgent)
        }
    }
}
```

The procedure *backtrack* is meant to locally correct inconsistencies. First, a new NoGood constraint has to be generated. Generating a NoGood is done by checking all assignments that are present into the agent agent view. If one of these is removed, and then the agent succeeds in choosing a new value for its variable, it means that such an assignment is wrong. Hence, that assignment is added to the NoGood constraint. After this phase, the new generated NoGood can be empty or not. If no assignment appears within that new constraint, then there is no solution for the DCSP. Otherwise, a NoGood message has to be sent to the lowest priority agent, and then its assignment has to be removed from agent view. Then, a final check of the agent view is done. JADEL implementation of such a procedure is another one-shot behaviour, whose code follows precisely the original pseudocode of the algorithm.

```
oneshot behaviour Backtrack
  for ABTAgent {
    do {
        var V = new HashMap<AID, Integer>(theAgent.agentview)
        var sortedVariablesList = V.keySet.sort
        V.remove(theAgent.AID)

        for(v : sortedVariablesList) {
            var removed = V.remove(v)
            if(theAgent.assignVariable(V)) V.put(v, removed)
        }

        if(V.isEmpty){
            activate behaviour SendNoSolution(theAgent)
        } else {
            activate behaviour SendNoGood(theAgent, V)

            theAgent.agentview.remove(V.keySet.max)

            activate behaviour CheckAgentView(theAgent)
        }
    }
}
```

Event Handlers. Others procedures specified in the original ABT pseudocode concern the reception of messages. When the agent receives an OK message, it has to update its agent view with that new information, then it must check if the new assignment is consistent with others in agent view. The reception of a message requires a cyclic behaviour, which waits cyclically for an event and checks if such an event is a message.

---

cyclic behaviour ReceiveOK **for** ABTAgent {

To ensure that such a message is the correct one, namely, an OK message, some conditions have to be specified. JADEL provides the construct on-when-do to handle this situation. The clause on identifies the type of event and eventually gives to it a name. If the event is a message, the clause when contains an expression that filters incoming messages, as follows.

---

on message msg
when {
    ontology is ABTOntology **and**
    performative is INFORM **and**
    content is OK
}

Conditions in when clause can be connected by logical connectives and, or, and they can be preceded by a not. They refer to the fields of the message, namely, ontology, performative, and content. Fields that are not relevant can be omitted, and multiple choices can be specified. For example a behaviour can accept REQUEST or QUERY_IF messages with performative is REQUEST or performative is QUERY_IF. The clause do is mandatory and contains the code of the action.

---

do {
    extract receivedOK **as** OK

    val a = receivedOK.assignment

    theAgent.agentview.replace(a.index, a.value)

    activate behaviour CheckAgentView(theAgent)
}

The content of the message is obtained by means of the JADEL expression extract- as, which manages all the needed implementation details and gives a name and a type to the content. Once the content of type OK of the message is obtained, its assignment is used to revise the agent view. Then, the behaviour CheckAgentView is activated.

Finally, the pseudocode of the procedure that manages the reception of a NoGood message is a cyclic behaviour for ABTAgent.

---

cyclic behaviour ReceiveNoGood **for** ABTAgent {

Checking if the event is a message, and then, if the message is actually a NoGood message, is done similarly to the OK reception, by using the clauses **on** and **when**, as shown in the following code.

```
on message msg
when {
    ontology is ABTOntology and
    performative is INFORM and
    content is NoGood
}
```

Inside the **do** body, the message content is extracted as a NoGood and it is recorded as a new constraint. We assume that the agent holds a set of constraints within the field `constraint` which is accessed by the agent instance `theAgent`.

```
do {
    extract receivedNoGood as NoGood

    val newConstraints = receivedNoGood . assignmentList

    theAgent . constraints . putAll ( newConstraints )
```

Then, if some constraints involve an agent which is not in the agent neighborhood, a request is sent to such an agent, in order to create a new link.

```
for (x : newConstraints . keySet ) {
    if (! theAgent . neighbors . contains (x)) {
        activate behaviour SendRequest ( theAgent , x )

        theAgent . neighbors . add ( x )
    }
}
```

Finally, the agent view must be checked, and if the previous value of the agent variable x_i remains unchanged, an OK message is sent.

```
var oldValue = theAgent . agentview . get ( theAgent . AID )

activate behaviour CheckAgentView ( theAgent )

if ( oldValue == theAgent . agentview . get ( theAgent . AID ) ) {
    activate behaviour SendOK ( theAgent )
}
```

3.2 Implementation of Savina Benchmarks

Other evaluations on JADEL are made by comparing it using the Savina benchmarks [29]. Savina is a benchmark suite to test actor libraries performances, and the source code of the thirty Savina benchmarks can be found at github.com/shamsimam/savina. For each benchmark, Savina provides an implementation by using the actor features of Akka [42], Functional-Java (www.functionaljava.org/), GPars (www.gpars.org/), Habanero-Java library [28], Jetlang (github.com/jetlang), Jumi (jumi.fi/actors.html), Lift (liftweb.net/api/26/api/#net.liftweb.actor.LiftActor), Scala [26], and Scalaz (github.com/scalaz). The thirty benchmarks that Savina provides are divided into *classic micro-benchmarks, concurrency benchmarks* and *parallelism benchmarks*.

Micro-benchmarks are simple benchmarks which test specific features of an actor library. For example, the classic `PingPong` benchmark measures the message passing overhead, while the `Counting` benchmark tests message delivery overhead. Concurrent benchmarks focus on classic concurrency problems, such as the dining philosophers, and they represent more realistic tests than micro-benchmarks. Finally, parallelism benchmarks exploit pipeline parallelism, phased computations, divide-and-conquer style parallelism, master-worker parallelism, and graph and tree navigation. In [29], the scope and the characteristics of each benchmark are discussed, and some experimental results are shown. It is worth noting that Savina is a suite that helps testing actor-oriented solutions, and it does not consider agent-oriented features. Nevertheless, Savina benchmarks are also suitable to analyze some features of agent programming languages, such as concurrency and message passing. For this reason, in this paper we take a few benchmarks from those proposed by Savina, and re-implemented them in JADEL. Savina does not yet contains inter-languages comparisons. As a matter of fact, sources are written in Java and Scala, and all benchmarks shows almost the same code: the differences among them are due to the various actor implementations. Additional language comparisons could be useful to evaluate the elegance, the readability and the simplicity of a given solution, beside its performances. Only a few Savina micro benchmarks are considered here, namely, the `PingPong`, `ThreadRing`, `Counting`, `Big`, `Chameneos` benchmarks.

It is worth noting that Savina benchmarks are thought for actor-based systems, and thus are heavily based on message passing. JADEL ontologies help in managing such task effectively. In the JADEL source code below, the simple ontology used for implementing the `PingPong` example is shown. The commented parts are the identifiers of the message objects which Savina implementation defines and uses for messages.

```
ontology PingPongOntology {
    proposition Start  // PingPongConfig.StartMessage
    proposition Ping   // PingPongConfig.SendPingMessage
    proposition Pong   // PingPongConfig.SendPongMessage
    proposition Stop   // StopMessage
}
```

The `PingPong` classic example consists in the definition of two agents which uses such an ontology, exchanging N `Ping` and `Pong` messages, alternatively. In the following listing, the source code of the ping agent, i.e., the initiator agent, is shown.

```
agent PingAgent uses ontology PingPongOntology {
    var AID pongAgent

    on create {
        pongAgent = newAID(arguments.get(0) as String)

        activate behaviour WaitForStartOrPong(this,
                PingPongConfig.N)
    }
}
```

Then, the pong agent, i.e., the responder agent has the following source code in JADEL.

```
agent PongAgent uses ontology PingPongOntology {
    var AID pingAgent

    on create {
        pingAgent = newAID(arguments.get(0) as String)

        activate behaviour WaitForPingOrStop(this, 0)

        activate behaviour SendInformMsg(this, #[pingAgent],
                new Start)
    }
}
```

Similarly, **ThreadRing** agents are defined. In this benchmark, N agents exchange R **Ping** messages, and they are limited to communicate only with the next agent in the ring. As for the **PingPong** benchmark, an ontology which follows precisely the Savina structure of message object is defined. In this example, message content are predicates rather than propositions, because they need to carry information between involved agents.

```
ontology ThreadRingOntology {
    predicate Ping(integer left)  // ThreadRingConfig.PingMessage
    predicate Data(aid next)       // ThreadRingConfig.DataMessage
    predicate Exit(integer left)  // ExitMessage
}
```

The JADEL source code for agent definition is listed below.

```
agent ThreadRingAgent extends JadelBaseAgent
    uses ontology ThreadRingOnto {
    var int id
    var AID na

    on create {
        na = newAID(arguments.get(0) as String)
        id = arguments.get(1) as Integer

        activate behaviour WaitForMsg(this)

        if (id == ThreadRingConfig.N - 1) {
            activate behaviour SendInformMsg(this, #[na],
                    new Ping(ThreadRingConfig.R))
        }
    }
}
```

As an example of `cyclic` behaviour, the following code shows the reception of an increment message in the `Counting` example. In this example, a `Producer` agent sends N increment messages to a `Counter` one, which counts the number of arrived messages. When the counter agent received all messages, it must inform the other agent of the resulting value of its count. As we can see, the behaviour `WaitForMsg` is a `cyclic` behaviour, as in ABT event handlers, because it must wait for a message and repeat its action each time a message arrives. For this scope, the construct `on-when-do` is used, as follows.

```
cyclic behaviour WaitForMsg for Counter {
    on message msg
    when {
        content is Increment
    } do {
        theAgent.count = theAgent.count + 1

        if (theAgent.count >= CountingConfig.N) {
            activate behaviour SendInformMsg(theAgent,
                    #[theAgent.producerAgent],
                    new ResultingValue(theAgent.count))

            activate behaviour Delete(theAgent)
        }
    }
}
```

Other micro benchmarks are implemented in the same fashion, with

1. A definition for each different kind of agent involved, which activates needed behaviours in the start up phase of its life cycle by means of the `on-create` handler;

2. The definition of a number of `cyclic` behaviour whose purpose is to intercept messages and process the correct ones; and
3. The definition of an ontology which terms are equivalent to the Savina ones.

Hence, the methodology used in [15] for implementing the ABT algorithm is the common way to creating agents and multi-agent systems with JADEL, whether the example is a very simple one (e.g., the `PingPong`), or a more complex algorithm as ABT.

4   Experimental Results

Methods to evaluate DSLs can be found in, e.g., [20], which focuses on multi-agent systems. Other surveys, such as [31,34], highlight the main advantages of the use of DSLs.

The comparison between the ABT pseudocode and its JADEL implementation is done by defining some metrics, which help us to get an idea of JADEL advantages and disadvantages. Then, we compare JADEL code with an equivalent JADE code, measuring the amount of code written, and the percentage of agent-oriented features of such a code. Nevertheless, comparing a pseudocode with an actual implementation is a difficult task, due to the informal nature of the pseudocode, and the implicit technical details it hides. Moreover, pseudocodes from different authors may look different, depending on their syntax choices and their purposes. As far as we know, there are not standard methods for evaluating the closeness of a source code to a pseudocode, and its actual effectiveness in expressing the described algorithm. Hence, we limit our evaluation to the use case of JADEL shown in this paper: the ABT example presented in previous section.

The first consideration that is made in evaluating the JADEL implementation of ABT is that ABT pseudocode is presented by means of procedures and event handlers, with the aid of the keywords `when` and `if`. As a second consideration, the notation used inside the ABT pseudocode is the same of the DCSP formalization. As a matter of fact, there are *agentview* and *neighbors* sets, and assignments are denoted as (x_i, d_i), where x_i is the variable associated with the i-th agent, and $d_i \in D_i$. A message is identified according to its type and its content, i.e., $(OK, (x_i, d_i))$ for an OK message, or $(nogood, (x_i, V))$ for a NoGood. Such characteristics of ABT pseudocode allow us to talk about *similarity* between it and the JADEL source code. In fact, in JADEL, both procedures and event handlers are represented as behaviours of the agent. In particular, procedures are one-shot behaviours that define an auto-triggering actions, while event handlers are cyclic behaviours, each of them waits for the given event and then performs its action. Hence, we can associate each behaviour with a procedure or an event handler, and analyze each of them separately. Moreover, calls to procedures in ABT pseudocode translate into the activation of the corresponding behaviour in JADEL. Also, the sending of a message is done by activating a specific JADEL behaviour. Hence, we associate each *send* instruction in ABT pseudocode to that activation. The DCSP notation is used also in JADEL,

by means of the two maps, *theAgent.agentview* and *theAgent.neighbors*, and by
defining some ontology terms. As a matter of fact, terms OK and $NoGood$ are
predicates in a JADEL ontology, and they contain an assignment, and a list of
assignments, respectively. Each assignment consists in a *index* and a *value*, i.e.,
x_i and d_i, respectively. The domain D_i of a variable is defined once in the start-
up phase of the agent and it is never modified during the execution of its actions.
We associate ABT pseudocode notations with the respective JADEL notation
described above. Finally, the reception of a message is done by using the construct `on-when-do`, which is the corresponding of ABT pseudocode construct
when $received(\dots)$ **do**.

We will say, in the following, that a line of ABT pseudocode *corresponds*
to a line (or, a set of lines) of JADEL implementation, if it falls in one of the
previous cases. Then, for each line of ABT pseudocode, we measure the number
of the corresponding *Lines Of Code (LOC)* of JADEL implementation. The
absolute value of the difference between ABT lines and corresponding JADEL
LOC is used as a first, rough, distance. For example, in the reception of an
OK message, the first line of the pseudocode corresponds to the `on-when-do`
constructs to capture the correct event, and filter other messages that are not
complied with the expected structure, as follows.

```
on message msg
when {
    ontology is ABTOnto and
    performative is INFORM and
    content is OK
}
```

Moreover, the `extract-as` expression is used to obtain the message content.

```
extract receivedOK as OK
```

Hence, we can conclude that in this case there are six LOCs instead of one line
of the pseudocode. Thus, the distance is of five LOCs. Such a distance gives us
an idea of the amount of code which is necessary to translate pseudocode into
JADEL, in case of ABT example. A summary is shown in [15] where a count of
nested blocks also presented. ABT pseudocode and JADEL implementation do
not differ significantly in terms of nested blocks, and JADEL code often requires
one more level (the `do` block), but its structure is usually very similar to ABT
pseudocode.

The count of nested blocks makes more sense when JADEL code is compared
to the equivalent JADE one. Such an equivalent implementation is obtained
directly from the available JADEL compiler [7], which translates JADEL code
into Java and uses JADE APIs. In fact, JADEL entities translate into classes
which can extend JADE `Agent`, `CyclicBehaviour`, `OneShotBehaviour`, and
`Ontology` base classes, while JADEL event handlers translate into the correct
methods of JADE APIs, in order to obtain the desired result. JADE code is automatically generated from the JADEL one, and this means that the final code
may introduce some redundancy or overhead. For this reason, we also write a

JADE code that implements ABT algorithm directly. Nevertheless, this alternative implementation is as complex as JADE generated code, because of some implementation details that JADE requires.

A comparison between JADEL and JADE implementation is made in terms of amount of code, i.e., by counting the number of non-comment and non-blank LOCs of each entity, namely, the `ABTAgent`, the `ABTOntology`, and all the behaviours. Results are shown in [15]. In order to emphasize the advantage in using JADEL instead of JADE, the percentage of lines which contains agent-oriented features over the total number of LOCs is also shown in [15]. We define as agent-oriented features each reference to the agent world. For example, keywords `agent`, `behaviour`, `ontology` are agent-oriented features, but are also special expressions such as `activate-behaviour`. In JADE, agent-oriented features are simply the calls to the API. [15] shows that the JADEL implementation is far lighter than the JADE one, and that it is more dense in terms of agent-oriented features. Such measures can be viewed as an indication of simplicity of JADEL code with respect to JADE.

The comparison between the chosen Savina benchmarks and their JADEL implementation is done by using the metrics of LOCs, with some restrictions. As in the ABT case, JADEL code is also compared with an equivalent JADE code, in terms of amount of code written. The main problem here is the different structure of a JADEL implementation, developed by using the JADEL approach, and the structure of a benchmark in Savina. Savina and JADEL projects are analyzed in terms of file written, utilities, and base classes, and only relevant parts of the benchmarks are evaluated (for example, configuration files are not counted). For a deeper evaluation, JADEL ontologies and Savina message objects are treated separately.

Savina benchmarks are structured as follows. There is a Java-written file of configurations, where parameters are initialized and managed (e.g. the number of pings N for the `PingPong` example), and objects for message passing are implemented. There is also a Scala source code that contains the implementations of actors, a class which implements the benchmark, i.e., a class that manages the iteration and the cleanup phases of each test, and an entry point for the benchmark. Similarly, JADEL benchmarks are structured as follows. The configuration file is the same as Savina. There is a Java file that implements the benchmark and the entry point. The most important, there is a JADEL file which contains the agents that are used in such a benchmark, their behaviours and an ontology for agent communications. It is worth noting that ontology predicates, concepts and propositions completely substitute that objects in the configuration file which are used in Savina for message passing. So, the JADEL implementation uses only a part of such a file, for getting parameter values, and does not take advantage of message objects.

Then, all benchmarks share a common base of methods and utilities. In Savina, for each considered actor framework, an actor base class is implemented.

Table 1. Number of LOCs, for Scala, JADEL and JADE implementation of selected examples from Savina benchmark suite.

	JADEL	Scala	JADE (generated source)
Base	55	68	197
PingPong	62	73	295
ThreadRing	46	53	239
Counting	72	40	308
Big	90	76	407
Chameneos	121	112	574
Philosopher	108	64	503

In JADEL, a base agent with two behaviours is implemented. Finally, Scala, JADEL and JADE LOCs are calculated using the following rules:

1. Blank lines or comments are not counted;
2. Prints for debugging are not counted;
3. Regarding Savina benchmarks, actors implementations are counted and also the definition and implementation of Message objects;
4. Regarding JADEL, the agent, behaviour and ontology implementations are counted; and
5. Regarding JADE, all generated files are counted.

Each measurement of the Savina suite is done by considering the Scala actor implementation of the benchmark. In Table 1, LOCs of some examples are shown, namely, the PingPong, ThreadRing, Counting, Big, Chameneos and Philosopher benchmarks. Table 2 emphasize the fact that JADEL syntax for ontologies is very light and the number of LOCs for ontologies remains little for each example, as opposed to Savina message objects or JADE ontologies.

Table 2. Number of LOCs, for Scala, JADEL and JADE implementation of ontologies and messages.

	JADEL ontology	Scala message objects	JADE ontology
Base	4	0	37
PingPong	6	29	57
ThreadRing	5	30	94
Counting	5	3	62
Big	5	21	47
Chameneos	7	32	141
Philosopher	7	6	97

5   Conclusions

In this paper, a quantitative evaluation of the agent-oriented programming language JADEL was presented. First, AOP paradigm and agent programming languages were briefly recalled, and JADEL was shortly presented. Then, a direct translation of the ABT pseudocode was presented together with a lighter presentation of the implementation of selected Savina benchmarks in JADEL. Such implementations were finally used quantitatively assess the performance of JADEL using specific metrics intended to evaluate the conciseness of the language and possible performance overheads introduced.

Not all adopted metrics can be regarded as complete in every situation, because they cannot fully describe qualitative factors, such as readability, reusability or maintainability. In particular, some of them heavily depend on the type of pseudocode given. However, such measurements help us at evaluating the simplicity of written code and they gives us an idea of the expressiveness and effectiveness of the language. As a matter of fact, the distance of JADEL from pseudocode and from Scala source code is very small. This fact may help JADE developers in translating an idea of distributed algorithms into a working JADE multi-agent system. Moreover, the distance from pseudocode and the number of nested blocks is very high in the case of JADE. This is mainly due to the very high number of implementation details that hide behind JADEL code, and the structure itself of Java language and JADE APIs.

References

1. Bădică, C., Budimac, Z., Burkhard, H.D., Ivanovic, M.: Software agents: languages, tools, platforms. Comput. Sci. Inf. Syst. **8**(2), 255–298 (2011)
2. Bellifemine, F., Bergenti, F., Caire, G., Poggi, A.: JADE - a Java agent development framework. In: Bordini, R.H., Dastani, M., Dix, J., El Fallah Seghrouchni, A. (eds.) Multi-Agent Programming: Languages, Platforms and Applications. MASA, vol. 15, pp. 125–147. Springer, Boston (2005). https://doi.org/10.1007/0-387-26350-0_5
3. Bellifemine, F., Caire, G., Greenwood, D.: Developing Multi-agent Systems with JADE. Wiley Series in Agent Technology. Wiley, Chichester (2007)
4. Bergenti, F., Iotti, E., Monica, S., Poggi, A.: Interaction protocols in the JADEL programming language. In: Proceedings of the 6th International Workshop on Programming Based on Actors, Agents, and Decentralized Control (AGERE!) (2016)
5. Bergenti, F., Iotti, E., Poggi, A.: An outline of the use of transition systems to formalize JADE agents and multi-agent systems. Intelligenza Artificiale **9**(2), 149–161 (2015)
6. Bergenti, F., Monica, S.: Location-aware social gaming with AMUSE. In: Trends in Practical Applications of Scalable Multi-Agent Systems, the PAAMS Collection (PAAMS 2016), pp. 36–47 (2016)
7. Bergenti, F.: An introduction to the JADEL programming language. In: Proceedings of the IEEE 26th International Conference on Tools with Artificial Intelligence (ICTAI), pp. 974–978. IEEE Press (2014)

8. Bergenti, F., Caire, G., Gotta, D.: Agents on the move: JADE for Android devices. In: Proceedings of the Workshop Dagli Oggetti Agli Agenti (WOA 2014). CEUR Workshop Proceedings, vol. 1260 (2014)

9. Bergenti, F., Caire, G., Gotta, D.: An overview of the AMUSE social gaming platform. In: Proceedings of the Workshop Dagli Oggetti agli Agenti (WOA 2013). CEUR Workshop Proceedings, vol. 1099 (2013)

10. Bergenti, F., Caire, G., Gotta, D.: Large-scale network and service management with WANTS. In: Industrial Agents: Emerging Applications of Software Agents in Industry, pp. 231–246. Elsevier (2015)

11. Bergenti, F., Franchi, E., Poggi, A.: Agent-based social networks for enterprise collaboration. In: Proceedings of the 20th International Conference on Enabling Technologies: Infrastructure for Collaborative Enterprises (WETICE 2011). IEEE Press (2011)

12. Bergenti, F., Franchi, E., Poggi, A.: Agent-based interpretations of classic network models. Comput. Math. Organ. Theory **19**(2), 105–127 (2013)

13. Bergenti, F., Gleizes, M.P., Zambonelli, F. (eds.): Methodologies and Software Engineering for Agent Systems: The Agent-Oriented Software Engineering Handbook. Springer, New York (2004). https://doi.org/10.1007/b116049

14. Bergenti, F., Iotti, E., Monica, S., Poggi, A.: A case study of the JADEL programming language. In: Proceedings of the Workshop Dagli Oggetti agli Agenti (WOA 2016). CEUR Workshop Proceedings, vol. 1664, pp. 85–90 (2016)

15. Bergenti, F., Iotti, E., Monica, S., Poggi, A.: A comparison between asynchronous backtracking pseudocode and its JADEL implementation. In: Proceedings of the 9th International Conference on Agents and Artificial Intelligence (ICAART), vol. 2, pp. 250–258. ScitePress (2017)

16. Bergenti, F., Iotti, E., Poggi, A.: Outline of a formalization of JADE multi-agents system. In: Proceedings of the Workshop Dagli Oggetti agli Agenti (WOA 2015). CEUR Workshop Proceedings, vol. 1382, pp. 123–128 (2015)

17. Bergenti, F., Iotti, E., Poggi, A.: Core features of an agent-oriented domain-specific language for JADE agents. In: de la Prieta, F., et al. (eds.) Trends in Practical Applications of Scalable Multi-Agent Systems, the PAAMS Collection. AISC, vol. 473, pp. 213–224. Springer, Cham (2016). https://doi.org/10.1007/978-3-319-40159-1_18

18. Bordini, R.H., Braubach, L., Dastani, M., Seghrouchni, A.E.F., Gomez-Sanz, J.J., Leite, J., O'Hare, G., Pokahr, A., Ricci, A.: A survey of programming languages and platforms for multi-agent systems. Informatica **30**(1), 33–44 (2006)

19. Braubach, L., Pokahr, A., Lamersdorf, W.: Jadex: a BDI-agent system combining middleware and reasoning. In: Unland, R., Calisti, M., Klusch, M. (eds.) Software Agent-Based Applications, Platforms and Development Kits, pp. 143–168. Birkhäuser (2005)

20. Challenger, M., Kardas, G., Tekinerdogan, B.: A systematic approach to evaluating domain-specific modeling language environments for multi-agent systems. Software Qual. J. **24**(3), 755–795 (2016)

21. Challenger, M., Mernik, M., Kardas, G., Kosar, T.: Declarative specifications for the development of multi-agent systems. Comput. Stand. Interfaces **43**, 91–115 (2016)

22. Demirkol, S., Challenger, M., Getir, S., Kosar, T., Kardas, G., Mernik, M.: SEA_L: a domain-specific language for Semantic Web enabled multi-agent systems. In: Federated Conference on Computer Science and Information Systems (FedCSIS), pp. 1373–1380 (2012)

23. El Fallah-Seghrouchni, A., Suna, A.: CLAIM: a computational language for autonomous, intelligent and mobile agents. In: Dastani, M.M., Dix, J., El Fallah-Seghrouchni, A. (eds.) ProMAS 2003. LNCS (LNAI), vol. 3067, pp. 90–110. Springer, Heidelberg (2004). https://doi.org/10.1007/978-3-540-25936-7_5

24. Fisher, M.: A survey of concurrent MetateM — the language and its applications. In: Gabbay, D.M., Ohlbach, H.J. (eds.) ICTL 1994. LNCS, vol. 827, pp. 480–505. Springer, Heidelberg (1994). https://doi.org/10.1007/BFb0014005

25. Foundation for Intelligent Physical Agents: FIPA specifications, multi-agents system standard specifications (2002). http://www.fipa.org/specifications

26. Haller, P., Odersky, M.: Scala actors: unifying thread-based and event-based programming. Theoret. Comput. Sci. **410**(2), 202–220 (2009)

27. Hindriks, K.V., De Boer, F.S., Van der Hoek, W., Meyer, J.J.C.: Agent programming in 3APL. Auton. Agent. Multi-Agent Syst. **2**(4), 357–401 (1999)

28. Imam, S.M., Sarkar, V.: Habanero-Java library: a Java 8 framework for multicore programming. In: Proceedings of the 2014 International Conference on Principles and Practices of Programming on the Java Platform: Virtual Machines, Languages, and Tools (PPPJ 2014), pp. 75–86. ACM (2014)

29. Imam, S.M., Sarkar, V.: Savina - an actor benchmark suite: enabling empirical evaluation of actor libraries. In: Proceedings of the 4th International Workshop on Programming based on Actors, Agents & Decentralized Control (AGERE!), pp. 67–80. ACM (2014)

30. Kravari, K., Bassiliades, N.: A survey of agent platforms. J. Artif. Soc. Soc. Simul. **18**(1), 11 (2015)

31. Mernik, M., Heering, J., Sloane, A.M.: When and how to develop domain-specific languages. ACM Comput. Surv. (CSUR) **37**(4), 316–344 (2005)

32. Monica, S., Bergenti, F.: Location-aware JADE agents in indoor scenarios. In: Proceedings of the Workshop Dagli Oggetti agli Agenti (WOA 2015). CEUR Workshop Proceedings, vol. 1382, pp. 103–108 (2015)

33. Monica, S., Bergenti, F.: A comparison of accurate indoor localization of static targets via WiFi and UWB Ranging. In: de la Prieta, F., et al. (eds.) Trends in Practical Applications of Scalable Multi-Agent Systems, the PAAMS Collection. AISC, vol. 473, pp. 111–123. Springer, Cham (2016). https://doi.org/10.1007/978-3-319-40159-1_9

34. Oliveira, N., Pereira, M.J., Henriques, P., Cruz, D.: Domain specific languages: a theoretical survey. In: INFORUM 2009 Simpósio de Informática. Faculdade de Ciências da Universidade de Lisboa (2009)

35. Poggi, A., Bergenti, F.: Developing smart emergency applications with multi-agent systems. Int. J. E-Health Med. Commun. **1**(4), 1–13 (2010)

36. Rao, A.S.: AgentSpeak(L): BDI agents speak out in a logical computable language. In: Van de Velde, W., Perram, J.W. (eds.) MAAMAW 1996. LNCS, vol. 1038, pp. 42–55. Springer, Heidelberg (1996). https://doi.org/10.1007/BFb0031845

37. Rodriguez, S., Gaud, N., Galland, S.: SARL: a general-purpose agent-oriented programming language. In: Proceedings of the IEEE/WIC/ACM International Joint Conferences of Web Intelligence (WI) and Intelligent Agent Technologies (IAT), vol. 3, pp. 103–110. IEEE Press (2014)

38. Shoham, Y.: AGENT-0: a simple agent language and its interpreter. In: Proceedings of the 9th National Conference on Artificial Intelligence (AAAI), vol. 91, pp. 704–709 (1991)

39. Shoham, Y.: Agent-oriented programming. Artif. Intell. **60**(1), 51–92 (1993)

40. Shoham, Y.: An overview of agent-oriented programming. In: Bradshaw, J. (ed.) Software Agents, vol. 4, pp. 271–290. MIT Press (1997)

41. Winikoff, M.: JACK intelligent agents: An industrial strength platform. In: Bordini, R.H., Dastani, M., Dix, J., El Fallah, S.A. (eds.) Multi-Agent Programming. MASA, vol. 15, pp. 175–193. Springer, Boston (2005). https://doi.org/10.1007/0-387-26350-0_7
42. Wyatt, D.: Akka concurrency. Artima Incorporation (2013)
43. Yokoo, M., Durfee, E.H., Ishida, T., Kuwabara, K.: The distributed constraint satisfaction problem: formalization and algorithms. IEEE Trans. Knowl. Data Eng. **10**(5), 673–685 (1998)
44. Yokoo, M., Hirayama, K.: Algorithms for distributed constraint satisfaction: a review. Auton. Agent. Multi-Agent Syst. **3**(2), 185–207 (2000)
45. Yokoo, M., Ishida, T., Durfee, E.H., Kuwabara, K.: Distributed constraint satisfaction for formalizing distributed problem solving. In: Proceedings of the 12th International Conference on Distributed Computing Systems, pp. 614–621. IEEE Press (1992)

Analogical Reasoning in Clinical Practice with Description Logic $\mathcal{ELH}$

Teeradaj Racharak[1,2(✉)] and Satoshi Tojo[2]

[1] School of Information, Computer, and Communication Technology,
Sirindhorn International Institute of Technology,
Thammasat University, Pathum Thani, Thailand
r.teeradaj@gmail.com
[2] School of Information Science, Japan Advanced Institute of Science
and Technology, Ishikawa, Japan
{racharak,tojo}@jaist.ac.jp

Abstract. Measuring concept similarity in ontologies is central to the functioning of many techniques such as ontology matching, ontology learning, and many related applications in the bio-medical domain. In this paper, we explore the relationship between clinical thought and analogy. More specifically, the paper formalizes the process of analogical reasoning in the phrase of diagnosis. That is, the phrase in which physicians have to reach an accurate explanation for the symptoms and signs found in a patient. Our approach is driven by the developing similarity measure in Description Logics called sim^π. Finally, the paper relates the approach to others and discuss future directions.

Keywords: Concept similarity measure · Description Logics
Analogical reasoning · Clinical reasoning · Expert system

1 Introduction and Motivation

Most Description Logics (DLs) are decidable fragments of first-order logic (FOL) [1] with clearly defined computational properties. DLs are the logical underpinnings of the DL flavor of OWL and OWL 2. The advantage of this close connection is that the extensive DLs literature and implementation experiences can be directly exploited by OWL tools. Description Logic $\mathcal{ELH}$ is also a fragment of the DLs that corresponds to the OWL EL profile, which is part of the W3C standard for an ontology language for the Semantic Web [2,3]. Our choice for the DL $\mathcal{ELH}$ is motivated by the fact that well-known medical ontologies, e.g. SNOMED CT [4,5] and the Gene Ontology [6], are engineered using this logic.

Concept similarity refers to human judgments of a degree to which a pair of concepts in question is similar. Measures of such concept similarity are computational techniques attempting to imitate the human judgments of concept similarity. Studies have shown that these contribute to many applications. For

© Springer International Publishing AG, part of Springer Nature 2018
J. van den Herik et al. (Eds.): ICAART 2017, LNAI 10839, pp. 179–204, 2018.
https://doi.org/10.1007/978-3-319-93581-2_10

instance, they are often used in ontology matching [7], ontology learning [8], and many related applications in the bio-medical domain [9].

Analogical reasoning makes use a kind of resemblance (e.g. semantic similarity) of one thing to another for assigning properties from one context to another. This kind of reasoning is used quite often by human beings in real-life situations, especially in clinical practice. The study in [10] shows that three basic ways of analogical reasoning are often used in clinical practice. Firstly, everyday medical thinking employs analogical reasoning in the process of clinical diagnosis. Secondly, it plays an important role in clinical teaching (both with theoretical and practical analogies). Lastly, analogical reasoning can be used in the teaching of analogical clinical reasoning. In this paper, we focus on the first way of their usage. That is, the phrase in which physicians have to reach an accurate explanation for the symptoms and signs found in a patient. It is worth noting that the study about medical thinking is received a lot of attention since the late '50s with two basic viewpoints, namely psychological study of medical expert reasoning and modeling of clinical reasoning (cf. [10,11] for further reading). This paper is going to focus on the second case where the reasoning framework is well-constructed from the psychological study of medical thinking.

To have an intuitive understanding of analogical reasoning in the phrase of diagnosis, let us take a look on the following case where a realistic medical thinking is exemplified in discovering diseases from a patient (cf. Example 1). In general, physicians may work with the clinical corpus, which can be represented by well-known bio-medical ontologies such as SNOMED CT [4,5] and the Gene Ontology [6]. We note that SNOMED CT terminologies are used throughout the paper to represent the clinical corpus as for the exemplification.

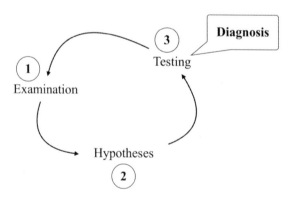

Fig. 1. The standard hypothetic-deductive model applied to diagnosis.

Example 1. Physician A is about to use the standard hypothetic-deductive model (cf. Fig. 1) in order to arrive at an explanation of the patient's signs and symptoms. Assume that, at the stage of examination, physician A finds the aspiration oral feed as the relevant sign. Considering physician A's theoretical knowledge, there could be several hypotheses to this sign. These are represented in the following terminologies, where Sign denotes the patient's current signs.

$$\begin{aligned}
\mathsf{Sign} &\sqsubseteq \mathsf{AspirationSyndromes} \\
&\sqcap \exists\mathsf{causativeAgent.DairyFoods} \\
\mathsf{AspirationOfMucus} &\sqsubseteq \mathsf{AspirationSyndromes} \\
&\sqcap \exists\mathsf{causativeAgent.Mucus} \\
\mathsf{AspirationOfMilk} &\sqsubseteq \mathsf{AspirationSyndromes} \sqcap \mathsf{InhalationOfLiquid} \\
&\sqcap \exists\mathsf{causativeAgent.Milk} \\
&\sqcap \exists\mathsf{associatedWith.Milk}
\end{aligned}$$

It is not difficult to see that both AOMu[1] and AOMi are related to Sign at certain degrees. This suggests physician A to confirm or disconfirm her hypotheses by effective techniques for arriving an accurate explanation.

In this paper, we model the analogical thought used by medical experts in the phrase of diagnosis. Several attempts have been made to study analogical reasoning in clinical reasoning. For instance, the so-called approach *clinical correlation* in [12] can also be considered as a kind of analogical reasoning. However, they are still lack of well-defined logical models; hence, it is not clear how analogical reasoning in clinical practice can be implemented. This is an extended study of our proceeding paper [13] to the topic of analogical reasoning in clinical practice (cf. Sects. 3 and 5). Our formal approach is driven by the technique of semantic similarity in Description Logics (cf. Sect. 4), which is originally introduced in our proceeding papers [13,14]. Preliminaries, related work, and the conclusion are discussed in Sects. 2, 6, and 7, respectively.

2 Preliminaries

In this section, we review the basics of Description Logic (DL) $\mathcal{ELH}$ (cf. Subsect. 2.1) and types of reasoning used in clinical diagnosis (cf. Subsect. 2.2).

2.1 Description Logic $\mathcal{ELH}$

We assume countably infinite sets CN of concept names, RN of role names, and IN of individual names that are fixed and disjoint. The set of concept descriptions, or simply concepts, for a specific DL $\mathcal{L}$ is denoted by $\mathsf{Con}(\mathcal{L})$. The set $\mathsf{Con}(\mathcal{ELH})$ of all $\mathcal{ELH}$ concepts can be inductively defined by the following grammar:

$$\mathsf{Con}(\mathcal{ELH}) ::= A \mid \top \mid C \sqcap D \mid \exists r.C$$

where $\top$ denotes the *top concept*, $A \in \mathsf{CN}$, $r \in \mathsf{RN}$, and $C, D \in \mathsf{Con}(\mathcal{ELH})$. Conventionally, concept names are denoted by A and B, concept descriptions

---

[1] For the sake of succinctness, obvious abbreviations may be used without being stated.

are denoted by C and D, and role names are denoted by r and s, all possibly with subscripts.

A *terminology* or *TBox* $\mathcal{T}$ is a finite set of (possibly primitive) concept definitions and role hierarchy axioms, whose syntax is an expression of the form $(A \sqsubseteq D)$ $A \equiv D$, and $r \sqsubseteq s$, respectively. A TBox is called *unfoldable* if it contains at most one concept definition for each concept name in CN and does not contain cyclic dependencies. Concept names occurring on the left-hand side of a concept definition are called defined concept names (denoted by $\mathsf{CN}^{\mathsf{def}}$), all other concept names are primitive concept names (denoted by $\mathsf{CN}^{\mathsf{pri}}$). A primitive definition $A \sqsubseteq D$ can easily be transformed into a semantically equivalent full definitions $A \equiv X \sqcap D$ where X is a fresh concept name. When a TBox $\mathcal{T}$ is unfoldable, concept names can be expanded by exhaustively replacing all defined concept names by their definitions until only primitive concept names remain. Such concept names are called *fully expanded concept names*. Like primitive definitions, a role hierarchy axiom $r \sqsubseteq s$ can be transformed in to a semantically equivalent role definition $r \equiv t \sqcap s$ where t is a fresh role name. Role names occurring on the left-hand side of a role definition are called defined role names, denoted by $\mathsf{RN}^{\mathsf{def}}$. All others are primitive role names, collectively denoted by $\mathsf{RN}^{\mathsf{pri}}$. We also denote a set of all r's super roles by $\mathcal{R}_r = \{s \in \mathsf{RN} | r = s$ or $r_i \sqsubseteq r_{i+1} \in \mathcal{T}$ where $1 \leq i \leq n, r_1 = r, r_n = s\}$.

An *assertion* or *ABox* $\mathcal{A}$ is a finite set of concept assertions and role assertions whose syntax is an expression of the form $C(a)$ and $r(a,b)$ where $a,b \in \mathsf{IN}$, respectively. An ontology $\mathcal{O}$ consists of a TBox $\mathcal{T}$ and an ABox $\mathcal{A}$, i.e. $\mathcal{O} = \langle \mathcal{T}, \mathcal{A} \rangle$. However, some existing ontologies may omit an ABox $\mathcal{A}$ in practice.

An *interpretation* $\mathcal{I}$ is a pair $\mathcal{I} = \langle \Delta^{\mathcal{I}}, \cdot^{\mathcal{I}} \rangle$ where $\Delta^{\mathcal{I}}$ is a non-empty set representing the domain of the interpretation and $\cdot^{\mathcal{I}}$ is an interpretation function which assigns to every concept name A a set $A^{\mathcal{I}} \subseteq \Delta^{\mathcal{I}}$, and to every role name r a binary relation $r^{\mathcal{I}} \subseteq \Delta^{\mathcal{I}} \times \Delta^{\mathcal{I}}$. The interpretation function $\cdot^{\mathcal{I}}$ is inductively extended to $\mathcal{ELH}$ concepts in the usual manner:

$$\top^{\mathcal{I}} = \Delta^{\mathcal{I}}; \qquad (C \sqcap D)^{\mathcal{I}} = C^{\mathcal{I}} \cap D^{\mathcal{I}};$$
$$(\exists r.C)^{\mathcal{I}} = \{a \in \Delta^{\mathcal{I}} \mid \exists b \in \Delta^{\mathcal{I}} : (a,b) \in r^{\mathcal{I}} \wedge b \in C^{\mathcal{I}}\},$$

An interpretation $\mathcal{I}$ is said to be a *model* of a TBox $\mathcal{T}$ (in symbols, $\mathcal{I} \models \mathcal{T}$) if it satisfies all axioms in $\mathcal{T}$. $\mathcal{I}$ satisfies axioms $A \sqsubseteq C$, $A \equiv C$, and $r \sqsubseteq s$, respectively, if $A^{\mathcal{I}} \subseteq C^{\mathcal{I}}$, $A^{\mathcal{I}} = C^{\mathcal{I}}$, and $r^{\mathcal{I}} \subseteq s^{\mathcal{I}}$. Also, an interpretation $\mathcal{I}$ is said to be a *model* of an ABox $\mathcal{A}$ (in symbols, $\mathcal{I} \models \mathcal{A}$) if it satisfies all axioms in $\mathcal{A}$. $\mathcal{I}$ satisfies axioms $C(a)$ and $r(a,b)$ if $a^{\mathcal{I}} \in C^{\mathcal{I}}$ and $(a,b) \in r^{\mathcal{I}}$, respectively. Furthermore, an interpretation $\mathcal{I}$ is said to be a *model* of an ontology $\mathcal{O}$ if it satisfies all axioms in $\mathcal{T}$ and $\mathcal{A}$. An interpretation $\mathcal{I}_{\mathcal{A}}$ is called the *canonical interpretation* if:

1. $\Delta^{\mathcal{I}_{\mathcal{A}}}$ of $\mathcal{I}_{\mathcal{A}}$ consists of all individual names in $\mathcal{A}$;
2. $\forall A \in \mathsf{CN}$, we define $A^{\mathcal{I}_{\mathcal{A}}} = \{x \mid A(x) \in \mathcal{A}\}$; and
3. $\forall r \in \mathsf{RN}$, we define $r^{\mathcal{I}_{\mathcal{A}}} = \{(x,y) \mid r(x,y) \in \mathcal{A}\}$.

That is, the canonical interpretation $\mathcal{I}_{\mathcal{A}}$ is the interpretation which takes the set of ABox as the interpretation domain.

The main inference problem for $\mathcal{ELH}$ is the subsumption problem. That is, given $C, D \in \mathsf{Con}(\mathcal{ELH})$ and an ontology $\mathcal{O}$, C is *subsumed* by D w.r.t. $\mathcal{O}$ (in symbols, $C \sqsubseteq_{\mathcal{O}} D$) if $C^{\mathcal{I}} \subseteq D^{\mathcal{I}}$ for every model $\mathcal{I}$ of $\mathcal{O}$. Furthermore, C and D are *equivalent* w.r.t. $\mathcal{O}$ (in symbols, $C \equiv_{\mathcal{O}} D$) if $C \sqsubseteq_{\mathcal{O}} D$ and $D \sqsubseteq_{\mathcal{O}} C$. A much more interesting inference problem, which is based on concept subsumption, is the concept hierarchy. That is, let $\mathsf{CN}(\mathcal{O})$ be the concept names occurring in $\mathcal{O}$, the concept hierarchy of $\mathcal{O}$ is the most compact representation of the partial ordering $(\mathsf{CN}(\mathcal{O}), \sqsubseteq_{\mathcal{O}})$ induced by the subsumption relation w.r.t $\mathcal{O}$. When an ontology $\mathcal{O}$ is empty or is clear from the context, we omit to denote $\mathcal{O}$, i.e. $C \sqsubseteq D$ or $C \equiv D$. Furthermore, a more generalization of the concept equivalence is a *concept similarity measure under preference profile* (cf. Definition 7), which is originally introduced in [14].

2.2   Types of Reasoning in Clinical Diagnosis

The studies [10,15] show that there are four basic kinds of clinical reasoning. In this work, we focus on the forth one; however, we also do review its relationship to others which will be summarized briefly as follows.

1. Deterministic or deductive reasoning: This is the simplest form of reasoning in terms rules of the form *sign → disease*. However, real practices may be not based only on deductive rules but also in terms of probabilities with a very high degree. This variation is called quasi-deterministic reasoning.
2. Inductive reasoning: This form focuses on quantitative relationship between signs and diseases without explicitly stating the links. For example, the correlation between the risk of having lung cancer and the amount of smoked cigarettes is relevant; however, a probabilistic correlation cannot explicitly establish the link *smoking → lung_cancer* or vice versa.
3. Causal reasoning: This kind of reasoning underlies all kinds of clinical reasoning but may not present in an explicit way. This form assumes that everything has a cause; hence, it focuses on seeking for the cause of each sign in a patient. It may also represents metaphysical bonds between signs and diseases instead of explicitly stating a rule of correlation.
4. Analogical reasoning: This kind of reasoning can be understood as a kind of resemblance of one thing to another; thus, assigning properties from one context to another. As we will see later, *not only comparing the current case with a theoretical model of a certain disease, it also allows to compare the current case with previous patients along with the corpus.*

Physicians often mix several types of reasoning for their work. Indeed, physician's diagnoses are primarily based on the use of analogical reasoning and causal reasoning (cf. Sect. 3), and inductive reasoning and deterministic reasoning are secondary ones. Analogy also reveals itself as a powerful tool in the search of a possible explanation of unknown diseases. In such situations, physicians need to compare the patients with any cases that relate to it and start to work on that direction. This advantage inspires us to study on *modeling of medical analogical thinking* from the psychological study of medical expert reasoning [10].

3   Modeling Analogical Thought in Clinical Reasoning

The standard hypothetic-deductive model (cf. Fig. 1) suggests that every research should begin with the examination. That is, from gathering the data to arriving at a diagnosis. However, this is always not the case in the daily practice. Unskilled physicians and expert ones manage their data and hypotheses in a radically different ways. Unskilled physicians at first focus *specific signs*; and then, try to make a *potential hypothetical diagnosis*. On the other hand, expert physicians at first begin with the *overall situation* and its possible causes; an then, try to seek for *additional signs* for confirming the diagnosis. These two directions are recognized as *forward analogical reasoning* and *backward analogical reasoning*, respectively. Figure 2 depicts the approaches, where the dashed arrow represents the forward approach and the solid arrow represents the backward approach.

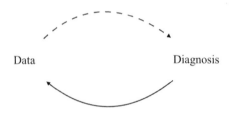

Data   Diagnosis

Fig. 2. Forward analogical reasoning (the dashed arrow) vs. backward analogical reasoning (the solid arrow).

Both ways of analogical reasoning are helpful for different circumstances. This section logically models these two approaches of clinical reasoning. To accomplish this, we formally define the fundamental framework for analogical reasoning and show later that the framework can be used for both the forward reasoning (cf. Subsect. 3.1) and the backward reasoning (cf. Subsect. 3.2).

We assume a formal language $\mathcal{S}$, a set $\mathcal{F}$ of sentences in $\mathcal{S}$, and a sentence $\mathcal{G}$ in $\mathcal{S}$. We also assume the strict consequence relation, i.e. $\mathcal{F} \vdash \mathcal{G}$, such that $\mathcal{F}$ are referred to as a set of facts whereas $\mathcal{G}$ is referred to as a goal. Unfortunately, when $\mathcal{G}$ cannot be strictly concluded from $\mathcal{F}$, rational agents may try to seek for additional evidences to support the conclusion and such conclusion may be *defeasible*. This process can be represented by the defeasible consequence relation, i.e. $\mathcal{F} \cup \mathcal{E} \mathbin{\vert\sim} \mathcal{G}$ where $\mathcal{E}$ represents a set of additional evidences.

Definition 1 (Basic Definition). *Let $\mathcal{F}$ be a set of facts in $\mathcal{S}$, $\mathcal{E}$ be a set of evidences in $\mathcal{S}$, and $\mathcal{G}$ be a goal in $\mathcal{S}$. Then, $\mathcal{G}$ is called an* analogical conclusion *from $\mathcal{F}$ and $\mathcal{E}$ if the following conditions are satisfied:*

- *$\mathcal{F} \nvdash \mathcal{G}$;*
- *$\mathcal{F} \cup \mathcal{E} \mathbin{\vert\sim} \mathcal{G}$; and*
- *$\mathcal{F} \cap \mathcal{E} \neq \emptyset$.*

The intuitive idea of Definition 1 is similar to other formal approaches (e.g. [16–19]) in which the attention is restricted to the situation that $\mathcal{G}$ is not deductively concluded from $\mathcal{F}$.

An important question which may arise from the basic definition is how $\mathcal{E}$ should be constructed? The key idea of analogy (cf. [20]) is *semantic similarity* or *semantic relatedness* and this can be served as an answer of the question.

A simple way to apply the basic definition in practice is using the following suggested inference rules[2], viz. a rule of defeasible modus ponens (MP) and a rule of defeasible analogical rule (AR) (adapted from the paper [21]).

$$\frac{F_1 \quad F_1 \Rightarrow G}{G} \text{ MP} \qquad\qquad \frac{F_1 \to G \quad F_2 \rightrightarrows F_1}{F_2 \Rightarrow G} \text{ AR}$$

We note that $F_1, F_2 \in \mathcal{F}$, $F_1 \neq F_2$, and $G \in \mathcal{G}$. Furthermore, $\to$ denotes a strict implication, $\Rightarrow$ denotes a defeasible implication, and $\rightrightarrows$ denotes semantic similarity or semantic relatedness between two sentences. We illustrate how the approach can be used to model analogical reasoning in Example 2.

Example 2. Assume that physician A's theoretical knowledge contains the link $sign_1 \to disease_1$ and the current patient she is diagnosing has a visible sign denoted as $sign'$. According to the basic definition, this situation may be modeled as $\mathcal{F} := \{sign_1 \to disease_1; sign'\}$[3]. It is obvious that $\mathcal{F} \not\vdash disease_1$. Hence, she has to seek for additional evidences $\mathcal{E}$ such that $\mathcal{F} \cup \mathcal{E} \hspace{0.5mm}\vdash\hspace{-2.2mm}\sim\hspace{1mm} disease_1$. Suppose that physician A knows $sign'$ is semantically similar to $sign_1$, i.e. $\mathcal{E} := \{sign' \rightrightarrows sign_1\}$. Therefore, $disease_1$ is an analogical conclusion based on her analogical thought.

If the consequence relations $\vdash$ and $\hspace{0.5mm}\vdash\hspace{-2.2mm}\sim$ are implemented in a sound and complete way, then they can be used in Definition 1 for the computation.

3.1 Forward Analogical Reasoning

As aforementioned, forward analogical reasoning relates to the process of making provisional hypothetical diagnoses from specific signs (cf. the dashed line of Fig. 2). This is often used by unskilled physicians or medical students. It is not difficult to see that this process is indeed a problem of finding $\mathcal{G}$ from a set $\mathcal{F}$ of facts and a set $\mathcal{E}$ of additional evidences.

Definition 2. *Let $\mathcal{F}$ and $\mathcal{E}$ be given. Then, a forward reasoner establishes $\mathcal{G}$ as a hypothetical diagnosis if $\mathcal{G}$ is an analogical conclusion from $\mathcal{F}$ and $\mathcal{E}$.*

It is worth observing that there can be finitely many choices for $\mathcal{G}$ as the cardinality of $\mathcal{E}$ may be infinite. This corresponds to the psychological study in medical thinking [10,22] that unskilled professionals tend to make a lot of hypothesis generation and testing. To trivialize the problem, we should restrict

---

[2] The precise interpretation of each symbol is given later.
[3] We separate each sentence with a semicolon.

our attention to only on a set of *sound* evidences. In other words, the key idea of forward analogical reasoning underlies on using sound evidences.

What could be sound evidences? The answer of this question may depend on preferences of a rational agent's choices. For example, ones may interest solely on the maximal degree of semantic similarity between two sentences (e.g. [19, 21]). Or, ones may allow the rational agent to manually set up a threshold with an agent-defined lower bound of similarity degree. These ideas can be used to filter out unsound hypothetical diagnoses.

3.2 Backward Analogical Reasoning

In contrast with forward analogical reasoning, backward analogical reasoning relates to the process of finding specific signs from plausible diagnoses (cf. the solid line of Fig. 2). This is often used by expert physicians. It is also not difficult to see that this process is indeed a problem of finding $\mathcal{E}$ from a set $\mathcal{F}$ of facts and a plausible diagnosis $\mathcal{G}$. Inferred $\mathcal{E}$ will be then used to confirm or dis-confirm their predicted diagnosis $\mathcal{G}$.

Definition 3. *Let $\mathcal{F}$ and $\mathcal{G}$ be given. Then, a backward reasoner establishes $\mathcal{E}$ as a hypothetical sign if $\mathcal{G}$ is an analogical conclusion from $\mathcal{F}$ and $\mathcal{E}$.*

It is also worth observing that there may be infinitely many choices for $\mathcal{E}$. We may trivialize the problem by restricting our attention only on *minimal* and *sound* evidences. More precisely, a set $\mathcal{E}$ is minimal iff there is no $\mathcal{E}' \subset \mathcal{E}$ such that the backward reasoner establishes $\mathcal{E}'$ as a hypothetical sign. Furthermore, handling the soundness of evidences can be done in the same fashion as forward analogical reasoning (cf. Subsect. 3.1).

Example 3. Assume that physician A's theoretical knowledge contains the link $sign_1 \rightarrow disease_1$. Based on her experience, she foresees that the current patient has $disease_1$. This situation may be modeled as $\mathcal{F} := \{sign_1 \rightarrow disease_1\}$ and $\mathcal{G} := disease_1$. It is obvious that $\mathcal{F} \not\vdash disease_1$. There, she has to seek for other invisible signs as evidences to support the prediction. Suppose that $\mathcal{E}_1 := \{sign_a \Rightarrow sign_1\}$, $\mathcal{E}_2 := \{sign_a \Rightarrow sign_1; a\}$, $\mathcal{E}_3 := \{sign_a \Rightarrow sign_1; a, aa\}$, ..., and $\mathcal{E}'_1 := \{sign_b \Rightarrow sign_1\}$, $\mathcal{E}'_2 := \{sign_b \Rightarrow sign_1; b\}$, $\mathcal{E}'_3 := \{sign_b \Rightarrow sign_1; b, bb\}^4$, ... It is not difficult to see that only $\mathcal{E}_1$ and $\mathcal{E}'_1$ are minimal. Hence, they are chosen as good candidates for hypothetical signs.

4   Semantic Similarity for Analogical Reasoning

It is important to notice that evidences used in analogical reasoning may involve measures of semantic similarity or semantic relatedness (cf. Examples 2 and 3). In this section, we present a conceptual notion of *concept similarity measure under the agent's preferences*, e.g. sim^π (originally introduced in [13, 14]) which can

---

[4] We note that a, aa, b, bb denote different arbitrary sentences in $\mathcal{S}$.

play an important role in supportive evidences for the analogical reasoning. As we will see, a rational agent may associate sentences with distinguished concept description. This way helps completely modeling analogical reasoning.

The subject of concept similarity in DLs has been widely studied. For instance, [23] develops two measures of concept similarity for DL $\mathcal{FL}_0$ and [24] develops a similarity measure for $\mathcal{EL}$ concepts (cf. Subsect. 6.2 for further discussion). In the following, we formally define the notion of concept similarity measure used in this paper and other work.

Definition 4. *Given $C, D \in \mathsf{Con}(\mathcal{L})$ be two concept descriptions for a specific DL $\mathcal{L}$. Then, a concept similarity measure w.r.t. a TBox $\mathcal{T}$ is a function $\sim_{\mathcal{T}}$: $\mathsf{Con}(\mathcal{L}) \times \mathsf{Con}(\mathcal{L}) \to [0, 1]$ such that $C \sim_{\mathcal{T}} D = 1$ iff $C \equiv_{\mathcal{T}} D$ (total similarity) and $C \sim_{\mathcal{T}} D = 0$ indicates total dissimilarity between C and D.*

When a TBox $\mathcal{T}$ is clear from the context, we simply write $C \sim D$. In addition, when two concepts are not equivalent, the degree of similarity varies depending not only on the objective factors (i.e. the concept descriptions) but also on the subjective factors (i.e. the agent's preferences). This observation is pointed out in [14,25] leading to the development of the so-called *preference profile* and *concept similarity measure under the agent's preferences* in DLs.

4.1 Preference Profile

We first introduced preference profile (denoted by π) in [25] as a collection of preferential elements in which the development of concept similarity measure should consider. Its first intuition is to model different forms of preferences (of an agent) based on concept names and role names. Measures adopted this notion are flexible to be tuned by an agent and can determine the similarity conformable to that agent's perception. In the original definition of preference profile [14,25], both i^c and i^r are mapped to $\mathbb{R}_{\geq 0}$, which is a minor error, and are fixed in [13].

Definition 5 (Preference Profile [13]). *Let $\mathsf{CN}^{\mathsf{pri}}(\mathcal{T})$, $\mathsf{RN}^{\mathsf{pri}}(\mathcal{T})$, and $\mathsf{RN}(\mathcal{T})$ be a set of primitive concept names occurring in $\mathcal{T}$, a set of primitive role names occurring in $\mathcal{T}$, and a set of role names occurring in $\mathcal{T}$, respectively. A* preference profile, *in symbol π, is a quintuple $\langle i^c, i^r, \mathfrak{s}^c, \mathfrak{s}^r, \eth \rangle$ where*

- $i^c : \mathsf{CN} \to [0, 2]$ *where $\mathsf{CN} \subseteq \mathsf{CN}^{\mathsf{pri}}(\mathcal{T})$ is called* primitive concept importance;
- $i^r : \mathsf{RN} \to [0, 2]$ *where $\mathsf{RN} \subseteq \mathsf{RN}(\mathcal{T})$ is called* role importance;
- $\mathfrak{s}^c : \mathsf{CN} \times \mathsf{CN} \to [0, 1]$ *where $\mathsf{CN} \subseteq \mathsf{CN}^{\mathsf{pri}}(\mathcal{T})$ is called* primitive concepts similarity;
- $\mathfrak{s}^r : \mathsf{RN} \times \mathsf{RN} \to [0, 1]$ *where $\mathsf{RN} \subseteq \mathsf{RN}^{\mathsf{pri}}(\mathcal{T})$ is called* primitive roles similarity; *and*
- $\eth : \mathsf{RN} \to [0, 1]$ *where $\mathsf{RN} \subseteq \mathsf{RN}(\mathcal{T})$ is called* role discount factor.

We discuss the interpretation of each above function in order. Firstly, for any $A \in \mathsf{CN}^{\mathsf{pri}}(\mathcal{T})$, $i^c(A) = 1$ captures an expression of normal importance on A, $i^c(A) > 1$ ($i^c(A) < 1$) indicates that A has higher (and lower, respectively) importance, and $i^c(A) = 0$ indicates that A has no importance to the agent. Secondly, we define the interpretation of i^r in the similar fashion as i^c for any $r \in RN(\mathcal{T})$. Thirdly, for any $a, b \in CN^{pri}(\mathcal{T})$, $s^c(A, B) = 1$ captures an expression of total similarity between A and B and $s^c(A, B) = 0$ captures an expression of total dissimilarity between A and B. Fourthly, the interpretation of s^r is defined in the similar fashion as s^c for any $r, s \in RN^{pri}(\mathcal{T})$. Lastly, for any $r \in RN(\mathcal{T})$, $\mathfrak{d}(r) = 1$ captures an expression of total importance on a role (over a corresponding nested concept) and $\mathfrak{d}(r) = 0$ captures an expression of total importance on a nested concept (over a corresponding role).

Definition 6 (Default Preference Profile [13]). *The default preference profile, in symbol π_0, is the quintuple $\langle i_0^c, i_0^r, s_0^c, s_0^r, \mathfrak{d}_0 \rangle$ where*

$$i_0^c(A) = 1 \text{ for all } A \in CN^{pri}(\mathcal{T}),$$
$$i_0^r(r) = 1 \text{ for all } r \in RN(\mathcal{T}),$$
$$s_0^c(A, B) = 0 \text{ for all } (A, B) \in CN^{pri}(\mathcal{T}) \times CN^{pri}(\mathcal{T}),$$
$$s_0^r(r, s) = 0 \text{ for all } (r, s) \in RN^{pri}(\mathcal{T}) \times RN^{pri}(\mathcal{T}),$$
$$\mathfrak{d}_0(r) = 0.4 \text{ for all } r \in RN(\mathcal{T}),$$

Let us also note that the value of $\mathfrak{d}_0$ determines how important the existential information should be considered by a measure in the default manner (see the interpretation of $\mathfrak{d}$) and is assigned the value 0.4 for its exemplification. This information is indeed dependent on an application domain and might be redefined. For instance, if $\mathfrak{d}_0$ is defined as 0.3, 0.4, 0.5, then $\exists part.Heart \sim_{\mathcal{T}}^{\pi} \exists part.Colon$ yields 0.3, 0.4, 0.5, respectively.

A continuous study on the strategies for tuning each preferential aspects is formally investigated in the paper [13]. These strategies help assigning values to each preferential aspect. In essence, a propagation for primitive importance $\mathfrak{I}^c$ assigns a corresponding weight to a group of related primitive concepts instead of individual assignment via i^c. Similarly, a propagation for role importance $\mathfrak{I}^r$ assigns a corresponding weight to a group of related roles instead of individual assignment via i^r. If an ABox $\mathcal{A}$ is presented, then we can induce the canonical interpretation $\mathcal{I}_\mathcal{A}$ from $\mathcal{A}$ to calculate primitive concepts similarity and primitive roles similarity for all possible primitive concept pairs and all possible primitive role pairs, respectively. It is worth noting that all these strategies may be helpful for the task of analogical reasoning with past experiences (cf. Subsect. 5.2 for a small discussion). Interested readers may refer to the paper [13] for detail.

4.2   Concept Similarity Measure Under the Agent's Preferences

As inspired by [25], the degree of similarity may vary by depending on subjective factors (e.g. the agent's preferences). Hence, the basic definition of concept similarity (cf. Definition 4) is modified to allow personalizing the measure. We formally define this new notion in the following.

Definition 7 ([14]). *Given a preference profile π, two concepts $C, D \in \mathsf{Con}(\mathcal{L})$, and a TBox $\mathcal{T}$, a concept similarity measure under preference profile w.r.t. a TBox $\mathcal{T}$ is a function $\overset{\pi}{\sim}_{\mathcal{T}} : \mathsf{Con}(\mathcal{L}) \times \mathsf{Con}(\mathcal{L}) \to [0,1]$. A function $\overset{\pi}{\sim}_{\mathcal{T}}$ is called preference invariance w.r.t equivalence if $C \equiv D \Leftrightarrow (C \overset{\pi}{\sim}_{\mathcal{T}} D = 1 \text{ for any } \pi)$.*

When a TBox $\mathcal{T}$ is clear from the context, we simply write $C \overset{\pi}{\sim} D$. It is not difficult to see that $\overset{\pi}{\sim}_{\mathcal{T}}$ can be used in a role of supportive evidences for analogical reasoning (cf. Definition 1 and the inference rule AR). As the goal we set ourselves at the onset of this paper is to explore the relationship between clinical thought and analogical reasoning, let us include the original definition of sim^{π} [13,14] – an instance of $\overset{\pi}{\sim}_{\mathcal{T}}$ for measuring similarity of $\mathcal{ELH}$ concepts.

Using sim^{π} requires that concept definitions in a TBox $\mathcal{T}$ must be fully expanded, i.e. for each defined concept name $A \in \mathsf{CN}^{\mathsf{def}}(\mathcal{T})$, such that $A \equiv D$, we simply replace A with D wherever it occurs in C and continue to recursively expand D. If A is of the form $A \sqsubseteq D$, then we replace A with $X \sqcap D$ such that X is a fresh concept wherever A occurs in C and recursively expand D. We note that X represents the primitiveness of A, i.e. the unspecified characteristics that differentiate it from D.

In order to consider all aspects of preference profile, we have presented a *total importance function* as $\hat{\mathsf{i}} : \mathsf{CN}^{\mathsf{pri}} \cup \mathsf{RN} \to [0,2]$ based on primitive concept importance and role importance.

$$
\hat{\mathsf{i}}(x) = \begin{cases} \mathsf{i}^{\mathsf{c}}(x) & \text{if } x \in \mathsf{CN}^{\mathsf{pri}} \text{ and } \mathsf{i}^{\mathsf{c}} \text{ is defined on } x \\ \mathsf{i}^{\mathsf{r}}(x) & \text{if } x \in \mathsf{RN} \text{ and } \mathsf{i}^{\mathsf{r}} \text{ is defined on } x \\ 1 & \text{otherwise} \end{cases} \tag{1}
$$

A *total similarity function* is also presented as $\hat{\mathsf{s}} : (\mathsf{CN}^{\mathsf{pri}} \times \mathsf{CN}^{\mathsf{pri}}) \cup (\mathsf{RN}^{\mathsf{pri}} \times \mathsf{RN}^{\mathsf{pri}}) \to [0,1]$ using primitive concepts similarity and primitive roles similarity.

$$
\hat{\mathsf{s}}(x,y) = \begin{cases} 1 & \text{if } x = y \\ \mathsf{s}^{\mathsf{c}}(x,y) & \text{if } (x,y) \in \mathsf{CN}^{\mathsf{pri}} \times \mathsf{CN}^{\mathsf{pri}} \\ & \text{and } \mathsf{s}^{\mathsf{c}} \text{ is defined on } (x,y) \\ \mathsf{s}^{\mathsf{r}}(x,y) & \text{if } (x,y) \in \mathsf{RN}^{\mathsf{pri}} \times \mathsf{RN}^{\mathsf{pri}} \\ & \text{and } \mathsf{s}^{\mathsf{r}} \text{ is defined on } (x,y) \\ 0 & \text{otherwise} \end{cases} \tag{2}
$$

Similarly, a *total role discount factor function* is presented in the following in term of a function $\hat{\partial} : \mathsf{RN} \to [0,1]$ based on role discount factor.

$$
\hat{\partial}(x) = \begin{cases} \partial(x) & \text{if } \partial \text{ is defined on } x \\ 0.4 & \text{otherwise} \end{cases} \tag{3}
$$

Let us note that the default value of Eqs. 1, 2 and 3 is set according to the default preference profile π_0 (Definition 6).

Let $C \in \mathsf{Con}(\mathcal{ELH})$ be a fully expanded concept to the form:

$$P_1 \sqcap \cdots \sqcap P_m \sqcap \exists r_1.C_1 \sqcap \cdots \sqcap \exists r_n.C_n$$

where $P_i \in \mathsf{CN}^{\mathsf{pri}}$, $r_j \in \mathsf{RN}$, $C_j \in \mathsf{Con}(\mathcal{ELH})$ in the same format, $1 \leq i \leq m$, and $1 \leq j \leq n$. The set $P_1, \ldots, P_m$ and the set $\exists r_1.C_1, \ldots, \exists r_n.C_n$ are denoted by $\mathcal{P}_C$ and $\mathcal{E}_C$, respectively. An $\mathcal{ELH}$ concept description can be structurally transformed into the corresponding $\mathcal{ELH}$ description tree. The root v_0 of the $\mathcal{ELH}$ description tree $\mathcal{T}_C$ has $\{P_1, \ldots, P_m\}$ as its label and has n outgoing edges, each labeled with r_j to a vertex v_j for $1 \leq j \leq n$. Then, a subtree with the root v_j is defined recursively relative to the concept C_j. Let $\pi = \langle \mathsf{i}^c, \mathsf{i}^r, \mathsf{s}^c, \mathsf{s}^r, \mathfrak{d} \rangle$ be a preference profile. The homomorphism degree under preference profile π can be formally defined as follows:

Definition 8 ([14]). *Let* $\mathbf{T}^{\mathcal{ELH}}$ *be a set of all* $\mathcal{ELH}$ *description trees and* $\mathcal{T}_C$, $\mathcal{T}_D \in \mathbf{T}^{\mathcal{ELH}}$ *corresponds to two* $\mathcal{ELH}$ *concept descriptions* C *and* D, *respectively. The* homomorphism degree under preference profile π *is a function* $\mathsf{hd}^\pi : \mathbf{T}^{\mathcal{ELH}} \times \mathbf{T}^{\mathcal{ELH}} \to [0,1]$ *defined inductively as follows:*

$$\mathsf{hd}^\pi(\mathcal{T}_D, \mathcal{T}_C) = \mu^\pi \cdot \mathsf{p\text{-}hd}^\pi(\mathcal{P}_D, \mathcal{P}_C) + (1 - \mu^\pi) \cdot \mathsf{e\text{-}set\text{-}hd}^\pi(\mathcal{E}_D, \mathcal{E}_C), \quad (4)$$

$$\text{where } \mu^\pi = \begin{cases} 1 & \text{if } \displaystyle\sum_{A \in \mathcal{P}_D} \hat{\mathsf{i}}(A) \\ & \quad + \displaystyle\sum_{\exists r.X \in \mathcal{E}_D} \hat{\mathsf{i}}(r) = 0 \\[2mm] \dfrac{\displaystyle\sum_{A \in \mathcal{P}_D} \hat{\mathsf{i}}(A)}{\displaystyle\sum_{A \in \mathcal{P}_D} \hat{\mathsf{i}}(A) + \displaystyle\sum_{\exists r.X \in \mathcal{E}_D} \hat{\mathsf{i}}(r)} & \text{otherwise;} \end{cases} \quad (5)$$

$$\mathsf{p\text{-}hd}^\pi(\mathcal{P}_D, \mathcal{P}_C) = \begin{cases} 1 & \text{if } \displaystyle\sum_{A \in \mathcal{P}_D} \hat{\mathsf{i}}(A) = 0 \\[2mm] 0 & \text{if } \displaystyle\sum_{A \in \mathcal{P}_D} \hat{\mathsf{i}}(A) \neq 0 \\ & \text{and } \displaystyle\sum_{B \in \mathcal{P}_C} \hat{\mathsf{i}}(B) = 0 \\[2mm] \dfrac{\displaystyle\sum_{A \in \mathcal{P}_D} \hat{\mathsf{i}}(A) \cdot \max\{\hat{\mathsf{s}}(A,B) : B \in \mathcal{P}_C\}}{\displaystyle\sum_{A \in \mathcal{P}_D} \hat{\mathsf{i}}(A)} & \text{otherwise;} \end{cases} \quad (6)$$

$$\mathsf{e\text{-}set\text{-}hd}^\pi(\mathcal{E}_D, \mathcal{E}_C) = \begin{cases} 1 & \text{if } \displaystyle\sum_{\exists r.X \in \mathcal{E}_D} \hat{\mathsf{i}}(r) = 0 \\[2mm] 0 & \text{if } \displaystyle\sum_{\exists r.X \in \mathcal{E}_D} \hat{\mathsf{i}}(r) \neq 0 \\ & \text{and} \\ & \displaystyle\sum_{\exists s.Y \in \mathcal{E}_C} \hat{\mathsf{i}}(s) = 0 \\[2mm] \dfrac{\displaystyle\sum_{\exists r.X \in \mathcal{E}_D} \hat{\mathsf{i}}(r) \cdot \max\{\mathsf{e\text{-}hd}^\pi(\exists r.X, \epsilon_j) : \epsilon_j \in \mathcal{E}_C\}}{\displaystyle\sum_{\exists r.X \in \mathcal{E}_D} \hat{\mathsf{i}}(r)} & \text{otherwise;} \end{cases} \quad (7)$$

where ϵ_j *is an existential restriction; and*

$$\mathsf{e\text{-}hd}^\pi(\exists r.X, \exists s.Y) = \gamma^\pi(\hat{\mathfrak{d}}(r) + (1 - \hat{\mathfrak{d}}(r)) \cdot \mathsf{hd}^\pi(\mathcal{T}_X, \mathcal{T}_Y)) \quad (8)$$

$$where \ \gamma^{\pi} = \begin{cases} 1 & if \ \sum_{r' \in \mathcal{R}_r} \hat{\mathsf{i}}(r') = 0 \\ \dfrac{\sum_{r' \in \mathcal{R}_r} \hat{\mathsf{i}}(r') \cdot \max\{\hat{\mathsf{s}}(r',s'):s' \in \mathcal{R}_s\}}{\sum_{r' \in \mathcal{R}_r} \hat{\mathsf{i}}(r')} & otherwise, \end{cases} \tag{9}$$

Intuitively, Eq. 4 is defined as the weighted sum of the degree under π of primitive concepts and the degree under π of matching edges. Equation 5 indicates the weight of primitive concept names w.r.t. the importance function. Equation 6 calculates the proportion of best similarity between primitive concept names. Similarly, Eq. 7 calculates the proportion of best similarity between existential information from Eqs. 8 and 9. Equation 8 calculates the degree of similarity between matching edges. Finally, Eq. 9 calculates the proportion of best similarity between role names.

Let C and D be fully expanded $\mathcal{ELH}$ concept names, $\mathcal{T}_C$ and $\mathcal{T}_D$ be the corresponding description trees, and $\pi = \langle \mathsf{i}^c, \mathsf{i}^r, \mathsf{s}^c, \mathsf{s}^r, \mathsf{d} \rangle$ be a preference profile. The following definition formally describes the $\mathcal{ELH}$ similarity degree under preference profile π.

Definition 9 ([14]). *The $\mathcal{ELH}$ similarity degree under preference profile π between C and D (denoted by $\mathsf{sim}^{\pi}(C, D)$) is defined as follows:*

$$\mathsf{sim}^{\pi}(C, D) = \frac{\mathsf{hd}^{\pi}(\mathcal{T}_C, \mathcal{T}_D) + \mathsf{hd}^{\pi}(\mathcal{T}_D, \mathcal{T}_C)}{2} \tag{10}$$

Intuitively, the degree of similarity under preference profile of two concepts is the average of the degree of having homomorphisms under preference profile in both directions. We note that $\mathsf{sim}^{\pi}(C, D) = 1$ indicates that C and D are total similarity under a particular π and $\mathsf{sim}^{\pi}(C, D) = 0$ indicates total dissimilarity between C and D under a particular π. Interested readers may refer to [14] for further discussion. In the following, we exemplify the calculation of sim^{π}.

Example 4 (Continuation of Example 1). Assume that the theoretical corpus is defined as follows:

$$
\begin{aligned}
\mathsf{AspirationOfMucus} &\sqsubseteq \mathsf{AspirationSyndromes} \\
&\quad \sqcap \exists \mathsf{causativeAgent.Mucus} \\
\mathsf{AspirationOfMilk} &\sqsubseteq \mathsf{AspirationSyndromes} \sqcap \mathsf{InhalationOfLiquid} \\
&\quad \sqcap \exists \mathsf{causativeAgent.Milk} \\
&\quad \sqcap \exists \mathsf{associatedWith.Milk} \\
\mathsf{Mucus} &\sqsubseteq \mathsf{BodySecretion} \\
\mathsf{BodySecretion} &\sqsubseteq \mathsf{BodySubstance} \\
\mathsf{BodySubstance} &\sqsubseteq \mathsf{Substance} \\
\mathsf{Milk} &\sqsubseteq \mathsf{DairyFoods} \\
\mathsf{DairyFoods} &\sqsubseteq \mathsf{FoodAllergen} \sqcap \mathsf{Foods} \\
\mathsf{FoodAllergen} &\sqsubseteq \mathsf{AllergenClass} \\
\mathsf{AllergenClass} &\sqsubseteq \mathsf{Substance} \\
\mathsf{causativeAgent} &\sqsubseteq \mathsf{associatedWith}
\end{aligned}
$$

Let us remind that the patient's sign is represented by the following definition.

$$\text{Sign} \sqsubseteq \text{AspirationSyndromes}$$
$$\sqcap \exists \text{causativeAgent.DairyFoods}$$

To use sim^π, concepts must be fully expanded. In the following, we present an semantically equivalent version of the theoretical corpus.

$$\text{AspirationOfMucus} \equiv A \sqcap \text{AspirationSyndromes}$$
$$\sqcap \exists \text{causativeAgent.}(C \sqcap D \sqcap E \sqcap \text{Substance})$$
$$\text{AspirationOfMilk} \equiv B \sqcap \text{AspirationSyndromes} \sqcap \text{InhalationOfLiquid}$$
$$\sqcap \exists \text{causativeAgent.}(F \sqcap G \sqcap H \sqcap I \sqcap \text{Substance} \sqcap \text{Foods})$$
$$\sqcap \exists \text{associatedWith.}(F \sqcap G \sqcap H \sqcap I \sqcap \text{Substance} \sqcap \text{Foods})$$
$$\text{Mucus} \equiv C \sqcap D \sqcap E \sqcap \text{Substance}$$
$$\text{BodySecretion} \equiv D \sqcap E \sqcap \text{Substance}$$
$$\text{BodySubstance} \equiv E \sqcap \text{Substance}$$
$$\text{Milk} \equiv F \sqcap G \sqcap H \sqcap I \sqcap \text{Substance} \sqcap \text{Foods}$$
$$\text{DairyFoods} \equiv G \sqcap H \sqcap I \sqcap \text{Substance} \sqcap \text{Foods}$$
$$\text{FoodAllergen} \equiv H \sqcap I \sqcap \text{Substance}$$
$$\text{AllergenClass} \equiv I \sqcap \text{Substance}$$

and $\text{Sign} \equiv J \sqcap \text{AspirationSyndromes} \sqcap \exists \text{causativeAgent.}(G \sqcap H \sqcap I \sqcap \text{Substance} \sqcap$ Foods), in which concepts A to J are fresh. We also note that $\mathcal{R}_{\text{causativeAgent}}$ denoting the super roles of causativeAgent is equal to $\{t \sqcap \text{associatedWith}\}$ where role t is also fresh.

Now, we are ready to calculate the similarity degrees. Let us take $\pi = \pi_0$ in this example.[5] We at first show the calculation of $\text{hd}^{\pi_0}(\mathcal{T}_{\text{Sign}}, \mathcal{T}_{\text{AOMi}})$ (see footnote 1).

$$\text{hd}^{\pi_0}(\mathcal{T}_{\text{Sign}}, \mathcal{T}_{\text{AOMi}}) =$$

$$= \left(\tfrac{2}{3}\right) \cdot \text{p-hd}^{\pi_0}(\mathcal{P}_{\text{Sign}}, \mathcal{P}_{\text{AOMi}}) + \left(\tfrac{1}{3}\right) \cdot \text{e-set-hd}^{\pi_0}(\mathcal{E}_{\text{Sign}}, \mathcal{E}_{\text{AOMi}})$$
$$= \left(\tfrac{2}{3}\right) \cdot \left(\tfrac{i(J) \cdot \max\{s(J,B), s(J,AS), s(J,IOL)\} + i(AS) \cdot \max\{s(AS,B), s(AS,AS), s(AS,IOL)\}}{i(J) + i(AS)}\right)$$
$$+ \left(\tfrac{1}{3}\right) \cdot \text{e-set-hd}^{\pi_0}(\mathcal{E}_{\text{Sign}}, \mathcal{E}_{\text{AOMi}})$$
$$= \left(\tfrac{2}{3}\right)\left(\tfrac{1 \cdot \max\{0,0,0\} + 1 \cdot \max\{0,1,0\}}{1+1}\right) + \left(\tfrac{1}{3}\right) \cdot \text{e-set-hd}^{\pi_0}(\mathcal{E}_{\text{Sign}}, \mathcal{E}_{\text{AOMi}})$$
$$= \left(\tfrac{2}{3}\right)\left(\tfrac{1}{2}\right) + \left(\tfrac{1}{3}\right) \left[\tfrac{i(cA) \cdot \max\{\text{e-hd}^{\pi_0}(\exists cA.(G \sqcap H \sqcap I \sqcap S \sqcap Fo), \exists cA.(F \sqcap G \sqcap H \sqcap I \sqcap S \sqcap Fo)), 0.5\}}{i(cA)}\right]$$
$$= \left(\tfrac{2}{3}\right)\left(\tfrac{1}{2}\right) + \left(\tfrac{1}{3}\right) \left[\tfrac{1 \cdot \max\{1, 0.5\}}{1}\right] \approx 0.667$$

Using the similar calculation, it yields $\text{hd}^{\pi_0}(\mathcal{T}_{\text{AOMi}}, \mathcal{T}_{\text{Sign}}) = 0.56$, $\text{hd}^{\pi_0}(\mathcal{T}_{\text{AOMu}}, \mathcal{T}_{\text{Sign}}) = 0.55$, and $\text{hd}^{\pi_0}(\mathcal{T}_{\text{Sign}}, \mathcal{T}_{\text{AOMu}}) \approx 0.512$. Hence, $\text{sim}^{\pi_0}(\text{Sign}, \text{AOMi}) \approx 0.614$ and $\text{sim}^{\pi_0}(\text{Sign}, \text{AOMu}) \approx 0.512$.

---

[5] We precisely explain specific intentions of taking different values for π in Sect. 5.

5   Doing Clinical Analogical Reasoning with sim$^\pi$

We can have a short summary from the previous sections. Analogical reasoning involves three components (cf. Definition 1), viz. a set $\mathcal{F}$ of facts, a set $\mathcal{E}$ of evidences, and a goal $\mathcal{G}$. Facts represent a real-life situation, such as unarguable observations and theoretical clinical knowledge. Evidences are additional information, such as similar degrees calculated from measures (e.g. sim$^\pi$) of concept similarity and presumption [21], used to support a goal which cannot be deductively inferred. In this section, we concretely show a simplistic way to employ the basic definition (cf. Definition 1) in clinical analogical reasoning.

To put things more concretely, let us assume the sentences in $\mathcal{L}$ are of the form: φ, $\varphi \to \psi$, $\varphi \Rightarrow \psi$, and $\varphi \rightrightarrows \psi$ where φ, ψ be constants and $\to, \Rightarrow, \rightrightarrows$ be strict implication, defeasible implication, and semantic similarity (as introduced in Sect. 3), respectively. Hence, the inference rules MP and AR can be used to make a derivation for analogical conclusions. Medical professionals may construct each sentence in $\mathcal{F}$ manually or employ automatic techniques to construct from an ontology. The following rule presents an automatic technique which translates $\mathcal{ELH}$ unfoldable TBox $\mathcal{T}$ to $\mathcal{F}$:

$$\text{If } C \equiv D \in \mathcal{T}, \text{ then } \underline{C} \to \underline{D} \in \mathcal{F} \text{ and } \underline{D} \to \underline{C} \in \mathcal{F} \tag{11}$$

where $\underline{\square}$ denotes the associated concept description $\square$ in $\mathcal{T}$. Indeed, the proposed rule can be used for translating $\mathcal{ELH}$ fully expanded concepts into sentences in $\mathcal{F}$. For instance, Sign is also translated into $\mathcal{F}$ (cf. Example 5).

Example 5 (Continuation of Example 4). It is not difficult to obtain the following sentences from the patient's sign and the theoretical knowledge. We note that $\underline{C} \leftrightarrow \underline{D}$ is an abbreviation of $\underline{C} \to \underline{D}$ and $\underline{D} \to \underline{C}$. The abbreviation is used only in the meta-representation.

$$\underline{\text{Sign}} \leftrightarrow J \sqcap \text{AspirationSyndromes}$$
$$\sqcap \exists\text{causativeAgent.}(G \sqcap H \sqcap I \sqcap \text{Substance} \sqcap \text{Foods})$$
$$\underline{\text{AspirationOfMucus}} \leftrightarrow A \sqcap \text{AspirationSyndromes}$$
$$\sqcap \exists\text{causativeAgent.}(C \sqcap D \sqcap E \sqcap \text{Substance})$$
$$\underline{\text{AspirationOfMilk}} \leftrightarrow B \sqcap \text{AspirationSyndromes} \sqcap \text{InhalationOfLiquid}$$
$$\sqcap \exists\text{causativeAgent.}(F \sqcap G \sqcap H \sqcap I \sqcap \text{Substance} \sqcap \text{Foods})$$
$$\sqcap \exists\text{associatedWith.}(F \sqcap G \sqcap H \sqcap I \sqcap \text{Substance} \sqcap \text{Foods})$$
$$\underline{\text{Mucus}} \leftrightarrow C \sqcap D \sqcap E \sqcap \text{Substance}$$
$$\underline{\text{BodySecretion}} \leftrightarrow D \sqcap E \sqcap \text{Substance}$$
$$\underline{\text{BodySubstance}} \leftrightarrow E \sqcap \text{Substance}$$
$$\underline{\text{Milk}} \leftrightarrow F \sqcap G \sqcap H \sqcap I \sqcap \text{Substance} \sqcap \text{Foods}$$
$$\underline{\text{DairyFoods}} \leftrightarrow G \sqcap H \sqcap I \sqcap \text{Substance} \sqcap \text{Foods}$$
$$\underline{\text{FoodAllergen}} \leftrightarrow H \sqcap I \sqcap \text{Substance}$$
$$\underline{\text{AllergenClass}} \leftrightarrow I \sqcap \text{Substance}$$

Measures of concept similarity, e.g. $\overset{\pi}{\sim}_{\mathcal{T}}$, is used for establishing sentences in $\mathcal{F}$. The following rule presents an automatic technique for this purpose:

$$\text{If } C \overset{\pi}{\sim}_{\mathcal{T}} D > 0, \text{ then } \underline{C} \rightrightarrows \underline{D} \in \mathcal{E} \tag{12}$$

where $\square$ denotes the associated concept description $\square$ in $\mathcal{T}$. For example, we have $J \sqcap \mathsf{AS} \sqcap \exists \mathsf{cA}.(G \sqcap H \sqcap I \sqcap S \sqcap \mathsf{Fo}) \;\; \rightrightarrows \;\; A \sqcap \mathsf{AS} \sqcap \exists \mathsf{cA}.(C \sqcap D \sqcap E \sqcap S)$ and $\overline{J \sqcap \mathsf{AS} \sqcap \exists \mathsf{cA}.(G \sqcap H \sqcap I \sqcap S \sqcap \mathsf{Fo})} \;\; \rightrightarrows \;\; B \sqcap \mathsf{AS} \sqcap \mathsf{IOL} \sqcap \exists \mathsf{cA}.(F \sqcap G \sqcap H \sqcap I \sqcap S$ $\sqcap \mathsf{Fo}) \sqcap \exists \mathsf{cW}.(F \sqcap G \sqcap H \sqcap I \sqcap S \sqcap \mathsf{Fo})$ as evidences in $\mathcal{E}$.

Within this formal approach, we have a systematic way to do clinical analogy. For example, it is not difficult to see that AOMi and AOMu are analogical conclusions based on physician A's analogical thought. They are plausible analogical conclusions in this example because both evidences support the derivation of AOMi and AOMu. However, which one should be more sound conclusion? As discussed in Subsects. 3.1 and 3.2, a rational agent may prefer to reason from maximal analogy or a satisfied threshold. Suppose that physician A uses the approach of maximal analogy, then AOMi is the only sound conclusion. This result can influence physician A's current decision to treat the patient in the same way as AOMi without doubt.

5.1   Analogy as Particularization

Analogy may be understood as a process of particularization, i.e. physicians compare the current case with the ideal case or the theoretical knowledge based on objective factors (e.g. the similarity of concepts' structure). We formally define this circumstance in the following definition.

Definition 10. *Let $\mathcal{F}$ be a set of facts in $\mathcal{S}$, $\mathcal{E}$ be a set of evidences in $\mathcal{S}$, and $\mathcal{G}$ be a goal in $\mathcal{S}$. Then, $\mathcal{G}$ is a* particularization *from $\mathcal{F}$ and $\mathcal{E}$ if the following conditions are satisfied:*

1. *$\mathcal{E}$ is constructed from the ideal case or the theoretical knowledge; and*
2. *$\mathcal{G}$ is an analogical conclusion from $\mathcal{F}$ and $\mathcal{E}$.*

The following theorem ensures that a process of particularization can be simulated from the use of sim^{π_0}. This trivializes an implementation of the process.

Theorem 1. *Let $\mathcal{T}$ be an unfoldable TBox and $\mathsf{CN}(\mathcal{T})$ be a set of concept names occurring in $\mathcal{T}$. Let $\mathcal{F}$ be constructed from $\mathcal{T}$ (e.g. Eq. 11) and $\mathcal{E}$ be constructed from the use of sim^{π_0} on $C, D \in \mathsf{CN}(\mathcal{T})$ (e.g. Eq. 12). Assume the soundness and completeness of $\vdash$ and $\mid\!\sim$. Then, $\mathcal{G}$ is particularized from $\mathcal{F}$ and $\mathcal{E}$ if $\mathcal{G}$ is an analogical conclusion from $\mathcal{F}$ and $\mathcal{E}$.*

Proof. *This is immediately hold if sim^{π_0} only measures from objective factors. Using Theorem 3.1 of [14], we know that $\mathsf{sim}^{\pi_0}(C, D) = \mathsf{sim}(C, D)$ for all $C, D \in \mathsf{CN}(\mathcal{T})$. This is trivially hold.* $\qquad\square$

The process of particularization may help unskilled physicians in the beginning; however, they may be not a good tool in the long run. It is very remarkable that diseases cannot be defined but are described in an idealized model of the patient (cf. [26], p. 26). Since every patient is unique, i.e. they have their own traits and have different historical patient records, then physicians have to take into account the patient's sign in accordance with his/her circumstance. For example, a young person and an old person are not the same case. We continue the discussion in Subsect. 5.2.

5.2   Analogy from Past Experiences

Since physicians are confronted with individual patients who exhibit different traits and peculiarities, it is important that physicians are capable of discerning relevant clinical signs from irrelevant features and assigning appropriate weights to the evidences. Basically, this can be seen as the physicians are relaxing the ideal case in appropriate ways. We formally define *relaxing* in general as follows.

Definition 11. *Let $\mathcal{T}$ be an unfoldable TBox, $\mathsf{CN}(\mathcal{T})$ be a set of concept names occurring in $\mathcal{T}$, Π be a countably infinite set of preference profile, and $\pi_1 \in \Pi$. Then, we call* relaxing *with π_1 if $C \overset{\pi_0}{\sim}_{\mathcal{T}} D < C \overset{\pi_1}{\sim}_{\mathcal{T}} D$ for some $C, D \in \mathsf{CN}(\mathcal{T})$.*

Definition 11 also implicitly says that physicians relax the ideal case to discover new similarity information. This corresponds to the fact that physicians discern relevant clinical signs differently from patients. We formally define analogy from past experiences in the following definition.

Definition 12. *Let $\mathcal{F}$ be a set of facts in $\mathcal{S}$, $\mathcal{E}$ be a set of evidences in $\mathcal{S}$, and $\mathcal{G}$ be a goal in $\mathcal{S}$. Assume Π be a countably infinite set of preference profile including profile settings used in the past diagnoses. Then, $\mathcal{G}$ is an* analogy from past experiences *from $\mathcal{F}$ and $\mathcal{E}$ if the following conditions are satisfied:*

1. *$\mathcal{E}$ is constructed from* relaxing *the ideal case or the theoretical knowledge with some $\pi \in \Pi$; and*
2. *$\mathcal{G}$ is an* analogical conclusion *from $\mathcal{F}$ and $\mathcal{E}$.*

Let Π be a countably infinite set of preference profile including profile settings used in the past diagnoses. The following theorem ensures that analogy from past experiences can be simulated from the use of sim^{π} for any $\pi \in \Pi$. This trivializes an implementation of the process.

Theorem 2. *Let $\mathcal{T}$ be an unfoldable TBox, $\mathsf{CN}(\mathcal{T})$ be a set of concept names occurring in $\mathcal{T}$, and Π be a countably infinite set of preference profile. Let $\mathcal{F}$ be constructed from $\mathcal{T}$ (e.g. Eq. 11) and $\mathcal{E}$ be constructed from the use of sim^{π} on $C, D \in \mathsf{CN}(\mathcal{T})$ relaxing with some $\pi \in \Pi$ (e.g. Eq. 12). Assume the soundness and completeness of $\vdash$ and $\vert\sim$. Then, $\mathcal{G}$ is an* analogy from past experiences *from $\mathcal{F}$ and $\mathcal{E}$ if $\mathcal{G}$ is an* analogical conclusion *from $\mathcal{F}$ and $\mathcal{E}$.*

Proof. *Fix any $\pi \in \Pi$ such that the ideal case is relaxed with π. It suffices to show that $\mathsf{sim}^{\pi}(C, D) > 0$ for some $C, D \in \mathsf{CN}(\mathcal{T})$. Using Definition 11, this is trivially hold.* ☐

Algorithm 1 describes a naive procedure of the forward analogical reasoning from past experiences based on Theorem 2. In practice, Π is always finite. It is worth observing that the time complexity is mainly contributed by an inference engine used to obtain analogical conclusions. A formal study of this is an issue of our future research (cf. Sect. 7).

Algorithm 1. Forward analogical reasoning from past experiences.

1: **function** GET-DIAGNOSES($\mathcal{T}$)
2:     $\mathcal{G}s := \emptyset$
3:     **for** $\pi \in \Pi$ **do**
4:         $\mathcal{F} := \{C \rightarrow D \mid C \equiv D \in \mathcal{T}\} \cup \{D \rightarrow C \mid C \equiv D \in \mathcal{T}\}$
5:         $\mathcal{E} := \{C \rightrightarrows D \mid \mathrm{sim}^{\pi}(C, D) > 0\}$
6:         **if** $\mathcal{G}$ is an analogical conclusion from $\mathcal{F}$ and $\mathcal{E}$ **then**
7:             $\mathcal{G}s := \mathcal{G}s \cup \mathcal{G}$
8:         **end if**
9:     **end for**
10:     **return** $\mathcal{G}s$
11: **end function**

It is worth noting that Algorithm 1 does not yet consider the aspect of sound evidences (cf. Subsect. 3.1). This can be improved by slightly modified the algorithm. In addition, the approach can be improved by considering only on relevant preference profile. Like other formal analogy-based approaches in other domains, e.g. the so-called case-based reasoning in legal reasoning (cf. the related work in [21] for useful discussion), decisions made for previous patients may influence a decision for the current patient. For instance, medical professionals may consider to apply used preference profile π for the current patient if the patient shares the same characteristics with prior cases, e.g. the same gender, the same age, and so on. Figure 3 illustrates an idea that historical patient records can be clustered regarding an applied preference profile. According to the figure, each dot represents each historical patient record. This illustration suggests that it is necessary to study on algorithmic procedures for clustering historical patient records and we leave this as our future study.

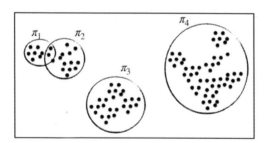

Fig. 3. Clustering of historical patient records with applied preference profile.

Ones may observe that doing analogy may require an effective way to fine tune the profile. It is worth noting that [13] explores strategies for effectively tuning the profile. For instance, medical professionals may employ the notion $\mathfrak{I}^c$ for collectively assigning weight to corresponding concepts (w.r.t the concept hierarchy). An example of this situation may be as follows. A physician is diagnosing a patient who visibly exhibits some risk factors to heart disease. In this situation, the physician may define $\mathfrak{J}^c(\mathsf{HeartDisease}) = 2$ instead of individually assigning $\mathfrak{i}^c(\mathsf{Pericarditis}) = 2$ and $\mathfrak{i}^c(\mathsf{Endocarditis}) = 2$ (assuming that the concept hierarchy is shown as in Fig. 4). Employing this kind of techniques is a shortcut for tuning preference profile. Interested readers may refer to [13] for other strategies, viz. $\mathfrak{J}^r$, $\mathfrak{S}^c$, and $\mathfrak{S}^r$, and further discussion.

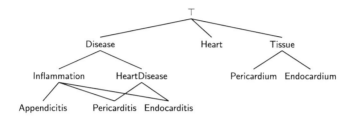

Fig. 4. An example of a concept hierarchy.

6   Related Work

As aforementioned, we study the logical model of analogical reasoning used in clinical practice. The model is powered by the measure of concept similarity called sim^π. According to this sense, this section reviews the approach to others with respect to two categories, viz. the use of analogy in reasoning (cf. Subsect. 6.1) and measures of concept similarity (cf. Subsect. 6.2).

6.1   The Use of Analogy in Reasoning

The use of analogy in reasoning is widely studied in many fields, such as argumentation study, logic, law, and clinical reasoning. While there is a diversity of approaches, there is some consensus, i.e. it is a form of non-deductive reasoning in which a conclusion is inferred based on similarity of two situations.

In the field of argumentation study, there are attempts on the development of reasoning schemes for analogical reasoning. Schemes are stereotypical non-deductive patterns of reasoning, consisting of a set of premises and a conclusion that is presumed to follow from the premises. It appears to us that there are basically two schemes for analogical reasoning. The first scheme developed by Walton et al. [27] enables to proceed from a source case to a target case based on the similarity between the source case to the target case as a supportive evidence. Use of Walton's scheme is evaluated by a set of critical questions. On the one hand, use of the scheme may visualize an understanding on how conclusion can be drawn from analogy. On the other hand, it is not difficult to observe that Walton's scheme can be represented by our formal approach.

Our intention is to gain the better understanding on analogical reasoning and how it can be implemented by any formal languages.

The second scheme derives a conclusion by comparing factors of two cases (e.g. [28,29]). For example, Mars revolves around the sun, collects light from the sun, revolves around its axis. This shares the same features with Earth. Hence, it is plausible to conclude that Mars may have living organisms like Earth. As inspired from this second scheme, work in the field of legal reasoning invents a well-known technique called *case-based reasoning*. Case-based reasoning uses dimensions and factors to realize if two cases are similar or different. Then, the most similar precedent case is then taken as the decision into the current case. HYPO [28] and CATO [29] are famous systems which use case-based reasoning.

There are substantial efforts of turning the results in philosophical study (e.g. [27,30]) into formal logics. For instance, [21,31] formulates Walton's scheme into a system of structured argumentation and answer set programming, respectively. A form of hypothetical reasoning is proposed in [19]. In this approach, similarity is expressed as an equality hypothesis and a goal-directed theorem prover is used to search relevant hypotheses. In [18,32], an analogical reasoning is considered as deductive reasoning by inserting a special rule to derive analogical conclusions. Another approach is [17] in which analogical reasoning is considered as partial identity of Horn clause logic interpretations.

In contrast with those logical models, we study a very general framework as a logical specification for analogical reasoning. While our modeling is inspired by clinical practice used in real-life situations, it can be applied for analogical reasoning used in other domains. Unlike most logical approaches, the specification does not restrict the attention only on maximizing the shared properties when it comes to multiple analogical conclusions. This corresponds to real-life diagnoses, where medical professionals may arrive at more than one hypotheses and they may have many reasons for arriving the conclusions. It is inevitable to notice that the ability of the formal approach depends on an applied measure of concept similarity. Fortunately, we show that applying with an instance of $\sim_{\mathcal{T}}^{\pi}$ (such as sim^{π}) can model many forms of analogical thought in clinical reasoning.

6.2   Measures of Concept Similarity

The subject of concept similarity is also widely studied in many fields, e.g. computer science, psychology, and linguistic literature. The techniques are roughly classified into four categories, viz. an ontology-based approach, an edit-distance-based approach, an information-theoretic approach, and a vector space approach.

Ontology-based approaches measure the degree of similarity based on ontology's characteristics. There is substantial work on methodology ranging from ontology hierarchy [33] to DLs-based theory [14,23,24,34–41]. A measure proposed by [33] relies on finding the most specific concept that subsumes both of the concepts being measured. On the other hand, approaches concerning DLs may be classified as two perspectives, viz. a structure-based approach [14,23,24,34–37,40,41] and an interpretation-based approach [38,39].

A structure-based approach is defined using the syntax of concepts being measured. A simple method is proposed in [40] for the DL $\mathcal{L}_0$ (i.e. no use of roles) and is known as Jaccard Index. Its extension to the DL $\mathcal{ELH}$ is proposed in [34]. This work also suggests desirable properties for concept similarity measures. Two measures for the DL $\mathcal{FL}_0$ are proposed in [23]. Both the skeptical and credulous measures are developed from the known structural characterization of subsumption through inclusion of regular languages. A similar approach based on the homomorphism structural characterization is originally proposed for the DL $\mathcal{EL}$ in [24]. Later, it is extended to the development of sim^π [13,14] for $\mathcal{ELH}$. Other developments of the structure-based measure are found in [35–37,41].

An interpretation-based measure uses only interpretations and cardinality to calculate the degree of similarity (e.g. [38,39]). It is also worth mentioning that there exist structure-based measures which use the canonical interpretation $\mathcal{I}_A$ in cases that primitive concepts are involved in the calculation (cf. [36,37]). In this sense, our proceeding work (sim^π) [13] which is a continuous study of [14] can also be classified in this category if used together with $\mathfrak{S}^{\mathfrak{c}}$ and $\mathfrak{S}^{\mathfrak{r}}$ since both strategies employ the canonical interpretation $\mathcal{I}_A$ for measuring similarity of primitive concept names and similarity of primitive role names, respectively.

The similarity between strings is often described as the edit distance [42]. It computes the changes to transform one string into one another. A changes is usually defined as insertion of a symbol, removal of a symbol, or replacement of a symbol with one another.

An information-theoretic entropy measure is immensely studied in the filed of natural language processing. This approach [43,44] describes that a concept can be defined by the member of a classed specified. The information of a class is defined as the probability of finding a use of the class. One drawback is the necessity of having a dataset to acquire the probability information.

A vector space approach is often used in information retrieval [45,46] and machine learning [47]. This approach represents a concept as a vector of features in a k-dimensional space and compute the similarity by vector-based measures (e.g. cosine similarity and Euclidean distance). Despite their usefulness in independent applicability to specific logic, the approach usually ignore to consider the real definitions in an ontology.

It is also worth observing that most of the above measures calculate the degree of concept similarity based on objective factors (e.g. the structure of a concept, the interpretation of a concept, and the probability). The study in [48] shows that real-life applications need more than one of measures in practice. This leads to an identification of agent preferences' aspects and is continuous on the development of sim^π [13,14].

7   Conclusion

We make contributions to the logical study of analogical reasoning. Indeed, we define a very general framework concerning a logical specification of analogical reasoning. Precisely speaking, three components constitute analogical thinking, viz. a set of facts, a set of additional evidences, and a goal. Analogical reasoning made by medical professionals is the main investigation cases of our study. As a result, we also show that many important forms of analogical thinking in clinical practice can be modeled based on our formal theory.

Analogy is dynamic in a sense that medical professionals need to compare the actual case with previous patients along with the theoretical corpus. This creates more subtle and complex comparisons. Our study shows that these situations can be trivialized with the use of $\sim_{\mathcal{T}}^{\pi}$ (e.g. sim^{π}). That is, sim^{π} and a set of prior preference profiles play important roles for the part of evidence set. In other words, there are four components of analogical reasoning, viz. a set of facts, sim^{π}, a set of preference profile, and a goal, for functioning both a process of particularization and analogy with past experiences. Furthermore, given an unfoldable TBox, our study also provides an automatic technique to construct the formal knowledge for reasoning. Let us also note that the relationship $sign \rightarrow disease$ used in the paper is only for the exemplification. Our formal language does not restrict to any specific kinds.

The psychological study of medical thinking [10, 26] finds that expertness of physicians influence the ability of thinking analogically. This is indeed related to variant forms of human expert reasoning. For example, unskilled physicians reason forward from data to diagnosis and expert physicians often reason backward from diagnosis to data. Furthermore. there is a relationship between the complexity of a disease and the usefulness of analogy. When a disease is unknown or circumstances are radically different from usual, physicians need to compare the patient with prior cases which have any resemblance to it and begin to work from that hypothetical diagnosis. Two primary aims of the formal theory are: (1) to reduce the gap of analogical use between unskilled physicians and expert physicians and (2) to help modeling a clinical expert system. The expert system should imitate the mental scheme of analogical reasoning and be able to suggest an answer which might be hardly unveiled by a human. Physicians may consider an applicability after the suggestion.

Our proceeding work [21, 31] study two different implementations of analogy reasoning. It is worth mentioning that both are specific forms of the general theory of analogical reasoning. On the one hand, [21] compute analogical conclusions using structured argumentation. On the other hand, [31] implements a system based on answer set programming to compute a conclusion. Hence, it is important to discover potential applications of different implementation methods. A formal study between the general theory of analogical reasoning and their different ways of the implementation is an issue of our future research.

There are a few directions for future work. Firstly, it appears to be natural next step to study on algorithmic procedures for each form of analogical reasoning formalized in this paper. For example, how should we do backward/forward analogical reasoning at the implementation level and what are their computational complexities? Secondly, we aim at carrying out applications for clinical diagnosis. Finally, it would be interesting to consider other methodologies of concept similarity measures and study how they contribute to analogical reasoning.

Concerning the approach of analogical reasoning with past experiences, it would be interesting to formally investigate a representation of the theoretical corpus for many possible worlds. That is, instead of having one global corpus, we may have many possible theoretical knowledge representing many possible definitions of the disease occurred in the past. This leads to a theoretical study of modal description logics [49] for clinical reasoning. This is left as a future task.

Acknowledgments. This research is part of the JAIST-NECTEC-SIIT dual doctoral degree program.

References

1. Baader, F., Calvanese, D., McGuinness, D.L., Nardi, D., Patel-Schneider, P.F.: The Description Logic Handbook: Theory, Implementation and Applications, 2nd edn. Cambridge University Press, New York (2010)
2. W3C Owl Working Group: OWL 2 web ontology language. Document overview, 2nd edn. W3C Recommendation, W3C, December 2012
3. Motik, B., Grau, B.C., Horrocks, I., Wu, Z., Fokoue, A., Lutz, C., et al.: Owl 2 web ontology language profiles, vol. 27, p. 61. W3C Recommendation (2009)
4. Benson, T., Grieve, G.: Principles of Health Interoperability: SNOMED CT, HL7 and FHIR. HITS. Springer, Cham (2016). https://doi.org/10.1007/978-3-319-30370-3
5. Spackman, K.: Managing clinical terminology hierarchies using algorithmic calculation of subsumption: experience with SNOMED-RT. J. Am. Med. Inform. Assoc. (2000)
6. Ashburner, M., Ball, C.A., Blake, J.A., Botstein, D., Butler, H., Cherry, J.M., Davis, A.P., Dolinski, K., Dwight, S.S., Eppig, J.T., et al.: Gene ontology: tool for the unification of biology. Nat. Genet. **25**(1), 25–29 (2000)
7. Euzenat, J., Shvaiko, P.: Ontology Matching. Springer, Heidelberg (2013). https://doi.org/10.1007/978-3-642-38721-0
8. Cohen, T., Widdows, D.: Empirical distributional semantics: methods and biomedical applications. J. Biomed. Inform. **42**(2), 390–405 (2009)
9. Pedersen, T., Pakhomov, S.V., Patwardhan, S., Chute, C.G.: Measures of semantic similarity and relatedness in the biomedical domain. J. Biomed. Inform. **40**(3), 288–299 (2007)
10. Guallart, N.: Analogical reasoning in clinical practice. In: Ribeiro, H.J. (ed.) Systematic Approaches to Argument by Analogy. AL, vol. 25, pp. 257–273. Springer, Cham (2014). https://doi.org/10.1007/978-3-319-06334-8_15
11. Price, C.: Computer-Based Diagnostic Systems. Practitioner Series, vol. 156. Springer, London (1999). https://doi.org/10.1007/978-1-4471-0535-0
12. Feinstein, A.R.: Clinical judgment (1967)
13. Racharak, T., Tojo, S.: Tuning agent's profile for similarity measure in description logic ELH. In: Proceedings of the 9th International Conference on Agents and Artificial Intelligence - Volume 2: ICAART, pp. 287–298 (2017)
14. Racharak, T., Suntisrivaraporn, B., Tojo, S.: sim$^\pi$: a concept similarity measure under an agent's preferences in description logic $\mathcal{ELH}$. In: Proceedings of the 8th International Conference on Agents and Artificial Intelligence, pp. 480–487 (2016)
15. Kassirer, J.P., Kopelman, R.I.: Learning clinical reasoning (1991)

16. Daley, R.P.: Towards the development of an analysis of learning algorithms. In: Jantke, K.P. (ed.) AII 1986. LNCS, vol. 265, pp. 1–18. Springer, Heidelberg (1987). https://doi.org/10.1007/3-540-18081-8_81

17. Haraguchi, M., Arikawa, S.: Reasoning by analogy as a partial identity between models. In: Jantke, K.P. (ed.) AII 1986. LNCS, vol. 265, pp. 61–87. Springer, Heidelberg (1987). https://doi.org/10.1007/3-540-18081-8_86

18. Greiner, R.: Learning by understanding analogies. In: Mitchell, T.M., Carbonell, J.G., Michalski, R.S. (eds.) Machine Learning. The Kluwer International Series in Engineering and Computer Science (Knowledge Representation, Learning and Expert Systems), vol. 12, pp. 81–84. Springer, Boston (1986). https://doi.org/10.1007/978-1-4613-2279-5_19

19. Goebel, R.: A sketch of analogy as reasoning with equality hypotheses. In: Jantke, K.P. (ed.) AII 1989. LNCS, vol. 397, pp. 243–253. Springer, Heidelberg (1989). https://doi.org/10.1007/3-540-51734-0_65

20. Walton, D.N.: Argumentation schemes for argument from analogy. In: Ribeiro, H.J. (ed.) Systematic Approaches to Argument by Analogy. AL, vol. 25, pp. 23–40. Springer, Cham (2014). https://doi.org/10.1007/978-3-319-06334-8_2

21. Racharak, T., Tojo, S., Hung, N.D., Boonkwan, P.: Argument-based logic programming for analogical reasoning. In: Kurahashi, S., Ohta, Y., Arai, S., Satoh, K., Bekki, D. (eds.) JSAI-isAI 2016. LNCS (LNAI), vol. 10247, pp. 253–269. Springer, Cham (2017). https://doi.org/10.1007/978-3-319-61572-1_17

22. Patel, V.L., Groen, G.J.: Knowledge based solution strategies in medical reasoning. Cogn. Sci. **10**(1), 91–116 (1986)

23. Racharak, T., Suntisrivaraporn, B.: Similarity measures for $\mathcal{FL}_0$ concept descriptions from an automata-theoretic point of view. In: Proceedings of the 6th International Conference of Information and Communication Technology for Embedded Systems (IC-ICTES), pp. 1–6, March 2015

24. Suntisrivaraporn, B.: A similarity measure for the description logic el with unfoldable terminologies. In: INCoS, pp. 408–413 (2013)

25. Racharak, T., Suntisrivaraporn, B., Tojo, S.: Identifying an agent's preferences toward similarity measures in description logics. In: Qi, G., Kozaki, K., Pan, J.Z., Yu, S. (eds.) JIST 2015. LNCS, vol. 9544, pp. 201–208. Springer, Cham (2016). https://doi.org/10.1007/978-3-319-31676-5_14

26. Cutler, P.: Problem Solving in Clinical Medicine: From Data to Diagnosis. Lippincott Williams & Wilkins, Philadelphia (1998)

27. Walton, D., Reed, C., Macagno, F.: Argumentation Schemes. Cambridge University Press, Cambridge (2008)

28. Ashley, K.: Case-based reasoning. In: Lodder, A.R., Oskamp, A. (eds.) Information Technology and Lawyers: Advanced Technology in the Legal Domain, from Challenges to Daily Routine, pp. 23–60. Springer, Dordrecht (2006). https://doi.org/10.1007/1-4020-4146-2_2

29. Aleven, V.: Teaching case-based argumentation through a model and examples, Ph.D. diss. University of Pittsburgh, Pittsburgh, Pennsylvania (1997)

30. Hofstadter, D., Mitchell, M.: Concepts, analogies, and creativity. In: Proceedings of CSCSI-88, pp. 94–101, June 1988

31. Racharak, T., Tojo, S., Hung, N.D., Boonkwan, P.: Combining answer set programming with description logics for analogical reasoning under an agent's preferences. In: Benferhat, S., Tabia, K., Ali, M. (eds.) IEA/AIE 2017. LNCS (LNAI), vol. 10351, pp. 306–316. Springer, Cham (2017). https://doi.org/10.1007/978-3-319-60045-1_33

32. Winston, P.H.: Learning and reasoning by analogy. Commun. ACM **23**(12), 689–703 (1980)
33. Wu, Z., Palmer, M.: Verbs semantics and lexical selection. In: Proceedings of the 32nd Annual Meeting on Association for Computational Linguistics, pp. 133–138. Association for Computational Linguistics (1994)
34. Lehmann, K., Turhan, A.-Y.: A framework for semantic-based similarity measures for $\mathcal{ELH}$-concepts. In: del Cerro, L.F., Herzig, A., Mengin, J. (eds.) JELIA 2012. LNCS (LNAI), vol. 7519, pp. 307–319. Springer, Heidelberg (2012). https://doi.org/10.1007/978-3-642-33353-8_24
35. Janowicz, K., Wilkes, M.: SIM-DLA: a novel semantic similarity measure for description logics reducing inter-concept to inter-instance similarity. In: Proceedings of the 6th European Semantic Web Conference on The Semantic Web: Research and Applications, pp. 353–367 (2009)
36. D'Amato, C., Fanizzi, N., Esposito, F.: A dissimilarity measure for ALC concept descriptions. In: Proceedings of the 2006 ACM Symposium on Applied Computing, pp. 1695–1699 (2006)
37. Fanizzi, N., D'Amato, C.: A similarity measure for the ALN description logic. In: Proceedings of CILC 2006 - Italian Conference on Computational Logic, pp. 26–27 (2006)
38. D'Amato, C., Fanizzi, N., Esposito, F.: A semantic similarity measure for expressive description logics. In: CoRR abs/0911.5043 (2009)
39. d'Amato, C., Staab, S., Fanizzi, N.: On the influence of description logics ontologies on conceptual similarity. In: Gangemi, A., Euzenat, J. (eds.) EKAW 2008. LNCS (LNAI), vol. 5268, pp. 48–63. Springer, Heidelberg (2008). https://doi.org/10.1007/978-3-540-87696-0_7
40. Jaccard, P.: Étude comparative de la distribution florale dans une portion des alpeset des jura. Bulletin de la Societe Vaudoise des Sciences Naturellese **37**, 547–579 (1901)
41. Tongphu, S., Suntisrivaraporn, B.: Algorithms for measuring similarity between ELH concept descriptions: a case study on SNOMED CT. J. Comput. Inform. **36**(4), 733–764 (2017)
42. Levenshtein, V.I.: Binary codes capable of correcting deletions, insertions, and reversals. Soviet Phys. Dokl. **10**, 707–710 (1966)
43. Resnik, P.: Using information content to evaluate semantic similarity in a taxonomy. In: Proceedings of the 14th International Joint Conference on Artificial Intelligence, pp. 448–453 (1995)
44. Resnik, P., et al.: Semantic similarity in a taxonomy: an information-based measure and its application to problems of ambiguity in natural language. J. Artif. Intell. Res. (JAIR) **11**, 95–130 (1999)
45. Baeza-Yates, R., Ribeiro-Neto, B., et al.: Modern Information Retrieval, vol. 463. ACM Press, New York (1999)
46. Patwardhan, S., Pedersen, T.: Using WordNet-based context vectors to estimate the semantic relatedness of concepts. In: Proceedings of the EACL 2006 Workshop Making Sense of Sense-Bringing Computational Linguistics and Psycholinguistics Together, Trento, vol. 1501, pp. 1–8 (2006)
47. Mitchell, T.M.: Machine Learning, vol. 45(37), pp. 870–877. McGraw Hill, Burr Ridge, IL (1997)

48. Bernstein, A., Kaufmann, E., Bürki, C., Klein, M.: How similar is it? Towards personalized similarity measures in ontologies. In: Ferstl, O.K., Sinz, E.J., Eckert, S., Isselhorst, T. (eds.) Wirtschaftsinformatik 2005, pp. 1347–1366. Physica-Verlag HD, Heidelberg (2005). https://doi.org/10.1007/3-7908-1624-8_71
49. Woltert, P., Zakharyaschev, M.: Modal description logics: modalizing roles. Fundam. Inform. **39**(4), 411–438 (1999)

Advanced User Interfaces for Semantic Annotation of Complex Relations in Text

Jaroslav Dytrych[(⊠)] and Pavel Smrz[(⊠)]

Faculty of Information Technology, Centre of Excellence IT4Innovations,
Brno University of Technology, Bozetechova 2, 61266 Brno, Czech Republic
{dytrych,smrz}@fit.vutbr.cz

Abstract. This paper deals with computer-assisted semantic annotation of text. It particularly focuses on the annotation of complex relations and linking of entities with highly ambiguous names. These tasks cannot be reliably accomplished by fully automatic methods today. Our research explores user interface features that can help the manual annotation process. We extend our original experiments published in [5] by a detailed analysis of advantages brought by the semantic filtering feature of our 4A annotation system. We also expand our user study on the annotation of highly ambiguous entities showing speed-ups brought by a presentation mode for entity candidates employing advanced disambiguation contexts.

Keywords: Computer-assisted tagging · Text annotation
Ambiguous entity names

1 Introduction

Semantic enrichment of text forms an initial step of various text analytics techniques that have been recently applied in brand reputation management, news recommendation, market research, and many other business domains. Commercial APIs such as IBM AlchemyLanguage[1], Cogito API[2], or Ontotext S4[3] automatize the task of semantic enrichment and enable annotating key entities and basic relations between them with an acceptable degree of precision. However, the quality of results achieved using fully automatic approaches varies significantly across annotation tasks and input data. It can be particularly low for complex and highly ambiguous cases [12,19]. The methods behind the most advanced tools usually employ machine-learning techniques which need training data. Consequently, they can be successfully applied for simple annotation tasks where the data is available but they fail for complex ones when there is not enough data to learn from. Manual annotation is necessary in such cases. The

---

[1] http://www.alchemyapi.com.

[2] http://cogitoapi.com.

[3] http://s4.ontotext.com.

© Springer International Publishing AG, part of Springer Nature 2018
J. van den Herik et al. (Eds.): ICAART 2017, LNAI 10839, pp. 205–221, 2018.
https://doi.org/10.1007/978-3-319-93581-2_11

research presented in this paper can be seen as a quest for an optimal way to support users in the manual annotation process.

Two approaches to manual annotation can be distinguished. The first one is represented by BRAT [18] – a general linguistic annotation editor that has been used to prepare various text annotation datasets. A key characteristics of this tool is a flat structure of basic annotations and a simple way of their presentation. For example, only semantic types of particular words or multi-word expressions are shown in the case of relation annotation in BRAT and it is not easy to solve ambiguities and to link entities to an external knowledge source. The tool is suitable for highly specialised tasks such as co-reference resolution or extraction of relations between biomedical entities which have unique names. On the other hand, knowledge base linking of ambiguous people names or complex hierarchical event annotation are not supported.

The second approach to computer-assisted manual annotation employs specialised semantic editors, plugins for existing tagging tools and web browser annotation extensions [3,6–8]. These systems are less suitable for general linguistic annotation tasks but they excel in semantic knowledge structuring (e.g., through a support of the RDF Schema) and name disambiguation. The 4A tool [16,17] primarily used in reported experiments also belongs to this category.

The high number of existing annotation systems contrasts with the fact that there are very few studies comparing particular features of the tools and discussing their suitability for specific tasks (some of them are briefly reviewed in Sect. 2 – Related work). To the best of our knowledge, no existing study compares design patterns employed in such tools that are relevant for complex annotation and disambiguation of highly ambiguous entities. This was the main purpose of the set of experiments conducted by our team and reported originally in [5].

The study compared three semantic annotation tools – GATE [4], RDFaCE [9], and 4A on the task of hierarchical annotation of complex relations. The annotation process consisted of selecting parts of a text corresponding to an event of a specific type, filling its attributes (slots) by entities and values mentioned in the text, and disambiguating the entities by linking them to a reference resource (mostly DBPedia/Wikipedia).

This paper extends the original results in two ways. Newly conducted experiments involve more annotators and bring new insights into interaction patterns observed in complex annotation tasks employing the 4A system. Experiments exploring benefits of 4A's semantic filtering functionality involve two consecutive tasks dealing with artwork authorship mentions and artistic influences expressed in texts.

Another set of experiments searched for an optimal amount of information necessary for reliable entity disambiguation. It showed that the commonly used practice of annotation tools asking users to disambiguate entities based just on a suggested type and a displayed URL leads to a poor quality of results. Extended experiments bring new results for the case of highly ambiguous entity names and demonstrate advantages of condensed entity views presenting task-tailored disambiguation attributes.

The rest of the paper is organized as follows: After Related work, Sect. 3 discusses high variability of text annotation tools and various factors that can influence comparison results. Research questions and experiments run to answer them are presented in Sect. 4. The last section summarizes reported results.

2   Related Work

As mentioned above, studies comparing user experiences with tools for semi-automatic text annotation are rare. The Knowledge Web project benchmarked 6 textual annotation tools considering various criteria including usability (installation, documentation, aesthetics. . .), accessibility (user interface features), and interoperability (platforms and formats) [11]. Most of the parameters are out of scope of our study but at least some of them are covered in a publically available feature matrix that we prepared to compare usage characteristics of annotation tools from the perspective of complex annotation tasks.[4]

Maynard [10] compares annotation tools from the perspective of a manual annotator, an annotation consumer, a corpus developer, and a system developer. Although the study is already 7 years old, most evaluation criteria are still valid. The study partially influenced our work.

Yee [21] motivates the implementation of CritLink – a tool enabling users to attach annotations to any location on any public web page – by a table summarizing shortcomings of existing web tools. As opposed to our approach, the comparison focuses on basic annotation tasks only and stresses technical aspects rather than the user experience.

Similarly to other studies, Reeve and Han [14] compare semantic annotation platforms focusing on the performance of background annotation suggestion components. As modern user interaction tools can freely change the back-ends and generate suggestions by a range of existing annotation systems, the work is relevant only from a historical perspective.

3   Factors Influencing Relevance of Tool Comparisons

When planning an experimental evaluation of semantic annotation frameworks, one has to take into account features significantly differing across the tools as well as varying aspects of the annotation process that can influence results of the studies.

Computer-assisted semantic annotation refers to a wide range of tasks. It can involve just a simple identification of entity mentions of few specific types in a text, but also full linking of potentially ambiguous entity names to a background knowledge base, annotation of complex hierarchical relations and their individual attributes. The domain of the text being annotated (e.g., biomedical v. general)

---

[4] http://knot.fit.vutbr.cz/annotations/comparison.html.

and its genre, register, or source (for example, news articles v. tweets) may also vary across annotation experiments. Consequently, the tasks can require particular approaches to text pre-processing and can imply different results of the automatic pre-annotation.

Datasets to be annotated do not necessarily correspond to a representative subset of relevant texts. They can focus on a chosen phenomenon and mix data accordingly. This variance can be demonstrated by differing nature of datasets prepared for previous annotation challenges. For example, the Short Text Track of the 2014 Entity Recognition and Disambiguation (ERD) Challenge[5] stressed limited contexts that naturally appear in web search queries from past TREC competitions. On the other hand, the SemEval-2015 Task 10[6] dealt with annotations relevant for sentiment analysis in microblog (Twitter) messages. The Entity Discovery and Linking (EDL) track at NIST TAC-KBP2015[7] then aimed at extracting named entity mentions, linking them to an existing Knowledge Base (KB) and clustering mentions for entities that do not have corresponding KB entries. Obviously, the degree of ambiguity of entities mentioned in annotated texts as well as proportions of occurrences corresponding to particular meanings can have a crucial impact on the speed and accuracy of the annotation process.

Experiments reported in this paper involve annotation of general web page texts (from the CommonCrawl corpus[8] – see below). Initial ones take random sentences based on trigger words (see Sect. 4.2 for details). Remaining experiments focus on entity linking tasks that are particularly difficult for automatic tools due to the ambiguity of names. We believe that making people annotate occurrences for which automatic tools often fail makes the scenario of manual annotation tasks more realistic. Sentences to be annotated are particularly selected to guarantee that there is at least one example of an occurrence corresponding to each potential meaning covered by the knowledge base. To evaluate a realistic setting, a part of the dataset is also formed by mentions not covered by the reference resources used.

Various aspects of annotation interfaces also influence experimental results. Some annotation tools aim at general applicability for semantic processes. Others are particularly tailored for paid-crowd annotation scenarios [2] so that they can be unsuitable for collaborative environments. Also, tools can be tied up with a particular annotation back-end or they can be only loosely coupled with a preferred annotator tool that can be easily changed or extended for specific tasks. Other features of annotation tools, especially those related to user interfaces and interaction patterns, are briefly discussed in the following section.

Skills, a current state of mind and motivation of users participating in experiments can also influence results. Measured quality and times always need to be interpreted with respect to these aspects. It can be expected that users with an experience in using a particular tool will better understand its user interface and

---

[5] http://web-ngram.research.microsoft.com/erd2014/.

[6] http://alt.qcri.org/semeval2015/task10/.

[7] http://nlp.cs.rpi.edu/kbp/2015/index.html.

[8] http://blog.commoncrawl.org.

will be able to achieve better results using the tool. Also, expertise in a domain in question can speed up the annotation process, especially the disambiguation of specialized entity mentions.

Experimental settings that can award quality over quantity or vice versa can lead to dramatically different times and amounts of annotation errors. Indeed, users in our experiments realized a trade-off between the time spent on each particular case and resulting quality (e.g., users' certainty that they considered enough context to correctly disambiguate an entity mention). While our users asked for preferences in this situation, this finding can be also expected in paid-for crowdsourcing settings that need to apply sophisticated quality control mechanisms to prevent annotators' temptation to cheat [20].

4   Annotation Experiments

4.1   Research Questions

Reported experiments aim at answering the following research questions:

1. How design choices of particular annotation tools impact the quality of results and the annotation time.
2. What quality the concept of semantic filtering brings to the annotation process.
3. To what extent the amount of information shown to disambiguators influences results.

Initial annotation experiments address the first question. They compare different user interfaces and interaction patterns as exemplified by three specific annotation systems. Various features that can influence annotation performance need to be considered. Some tools make no visible distinction between pre-annotations generated by a back-end automatic system and manual annotations entered by users. Other tools explicitly distinguish system suggestions from accepted or newly entered annotations. This can have an impact on the annotation consistency.

Underlying annotation patterns for events and other complex relations and their attributes vary across tools too. Advanced tools enable defining sophisticated templates and type constraints that control filling of event slots. Of course, systems differ in their actual application of the general approach and the way they implement it generally influences annotation performance aspects.

Various values entered by users often need to correspond to an entry in a controlled vocabulary or a list of potential items. An example of such a case is a URL linking an entity mention to a reference knowledge source. Tools support entering such values through autocomplete functions that can present not only the value to be entered, but also additional information that helps users to choose the right value. For example, the 4A client shows not only a URL link, but also full names and disambiguation contexts. The RDFaCE, on the other hand, autocompletes just URLs in this context.

The second set of experiments explores the role of semantic filtering. It is not easy to enter complex annotations, for example, interlinked hierarchical relations. Advanced mechanisms suggesting slot filling can make the process faster and more consistent. The 4A tool supports hierarchical annotation which highlights potential nested annotations if an upper-level type is known. Section 4.4 demonstrates that such an approach leads to significant speed-ups.

The last question mentioned above is covered by the final set of reported experiments. It is clear that the amount of information shown to the user and its form can influence speed and accuracy of annotation. If the displayed information is not sufficient for a decision, the user will have to search additional information. On the other hand, if a tool lets users read more than necessary, annotation speed decreases.

Most of the explored systems show just a URL and let the user explore it if she is not sure that the linked information corresponds to an expected one. This can speed up the annotation process but it can also make it error-prone. The 4A system enables filtering displayed information and fine-tuning the way it is shown. Detailed entity attributes can be folded and shown only if the user asks for them.

Without a system support, users are often unaware of ambiguity of some names. It causes no harm for frequent appearances of dominant senses. However, if a user is not an expert in a field where two or more potential links to an underlying resource can appear, she can easily confirm an incorrectly suggested link for an ambiguous name. The risk can be mitigated if the tools let users know about alternatives. The question is how an optimal setting for this function looks like – whether this should be a default behaviour or the system should notify the user only if an automatically computed confidence is smaller than a threshold or the difference between a suggested option and the second one is closer than a threshold. These aspects are discussed in the experiment too.

4.2   Data Preparation

Texts to be annotated in the experiments reported in the following subsections were chosen from general web data contained in the CommonCrawl corpus from December 2014.[9] First experiments dealt with general annotation of events. Text selection did not address any specific objective (in contrast to next experiments) so that contexts containing mentions of recognized named entities and a trigger word (verb) corresponding to artistic influences and travels and visits of people to various places were pre-selected. The data was then manually annotated by two authors of this paper, annotation disagreements were solved by choosing correct ones in clear cases and excluding few cases considered unclear.

The second set of experiments combine annotation of events with disambiguation of entity mentions. Consequently, paragraphs with sentences mentioning ambiguous names linked to the Wikipedia that contain a trigger verb indicating

---

[9] http://blog.commoncrawl.org/2014/12/.

a particular type of artistic influence relations were retrieved from the Com-monCrawl data and validated by the authors. Similarly to the data for the first experiments, the dataset consists of cases in which the pre-annotation process had led to a clear consensus between the annotators. Only 20 texts mentioning several artwork influence relations were used in the study.

Final experiments, looking into an optimal amount of displayed information, needed data containing ambiguous names with a proportional representation of two or more alternatives. Inspired by WikiLinks[10], we searched the Common-Crawl data for cases linking a name to two or more distinct Wikipedia URLs. To filter out potential interdependencies among various options and to enable focusing on key attributes in the first part of experiments, a majority of the pre-pared dataset consists of pairs of texts mentioning a name shared by two distinct entities. For example, the following sentences are included in the resulting data:

1. *Charles Thomson was a Patriot leader in Philadelphia during the American Revolution and the secretary of the Continental Congress (1774–1789) throughout its existence.*
2. *Charles Thomson's best known work is a satire of Sir Nicholas Serota, Director of the Tate gallery, and Tracey Emin, with whom he was friends in the 1980 s.*

Extended experiments focused on highly ambiguous names that could refer to tens of entities.

4.3   Comparing Tools

The aim of initial experiments was to compare advanced annotation editors in terms of their interaction patterns and user-interface features that can influence user experience and annotation performance. We were interested whether annotation results obtained by using particular tools will differ in the quality measured by their completeness and accuracy of types of entities filling slots of complex relations and their links to underlying resources (mostly DBPedia/Wikipedia). In addition, times to finish each experiment were measured for each user and then averaged per attribute annotated.

Employed tools represent different approaches to complex annotation tasks (see Fig. 1 for examples of event annotation views). The 4A system[11] pays a special attention to hierarchical annotations and potentially overlapping textual fragments. Users benefit from advanced annotation suggestions and an easy mechanism for entering correct attribute values by simply accepting or rejecting provided suggestions.

The RDFaCE editor[12] is similar to 4A in the way it annotates textual fragments and the fact it can be also deployed as a plugin for JavaScript WYSIWYG editor TinyMCE. It can pre-annotate texts too. On the other hand, there is no

---

[10] https://code.google.com/p/wiki-links/.

[11] http://knot.fit.vutbr.cz/annotations/.

[12] http://rdface.aksw.org/.

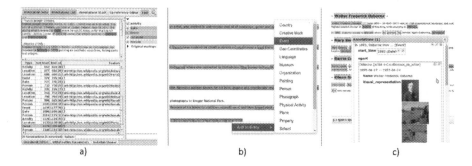

Fig. 1. Event annotation screens in (a) GATE, (b) RDFaCE, and (c) 4A.

visual distinction between a suggestion and a user annotation in RDFaCE. There is also no easy way to annotate two overlapping parts of a text with two separate events. Thus, testers were allowed to simplify their job and select whole sentences or even paragraphs as fragments corresponding to events.

Various existing GATE extensions and plugins were considered for the annotation experiment. Unfortunately, GATE Teamware[13] – a web-based collaborative text annotation framework which would be an obvious choice – does not currently provide good support for relation and co-reference annotation [1]. Similarly, simple question-based user interfaces generated by the GATE Crowdsourcing plugin[14] [2] would not be efficient for the complex hierarchical annotation tasks tested. Thus, our annotators used the standard GATE Developer[15] desktop interface, able to cope with the task at hand. Pre-annotations by back-end annotators were set the same way as in the other two tools. Users were instructed to perform an easier task of selecting event attributes and linking them to a reference resource first and then just selecting a text including all identified arguments and tagging it as an event.

As discussed in Sect. 3, it is very difficult to objectively compare semantic annotation tools from the user perspective. To minimize the danger of unfair comparison, six users that participated in the experiments had been selected to have no previous experience with neither the tools explored, nor the tasks that they used the tools for. They were 4 men and 2 women, PhD candidates or MSc. in computer science, aged 26–34. Every user spent about 20 min prior to the measured session familiarizing her-/himself with the tool to test while working on a specific part of the data, not included in the real testing set, yet containing all cases that would appear later during real testing (e.g., multiple values for attributes, two distinct events expressed in one sentence, suggestions that do not correspond to a correct sense, etc.) To make the comparison as fair as possible, the order in which users tested the tools was different for each user too.

---

[13] https://gate.ac.uk/teamware.

[14] https://gate.ac.uk/wiki/crowdsourcing.html.

[15] https://gate.ac.uk/family/developer.html.

Each user had about 40 min for annotation in each tool in the experiment. Three characteristics were collected. As summarized in Table 1, they included the amount of incorrect values entered, the number of misses – entities that were mentioned in the text but not associated with the event being annotated – and the average annotation time per event. Incorrect attributes involve all kinds of errors – incorrect selection of a textual fragment, blank or incorrect types, co-references or URLs linking entity mentions to a wrong entry in reference resources.

Table 1. Results of experiments comparing annotation tools.

Tool	Incorrect values	Missing values	Time per event
GATE	9.4%	8.3%	135 s
RDFaCE	8.7%	8.8%	193 s
4A	6.2%	5.6%	116 s

The overall high error rate (column "incorrect values") can be explained by rather strict comparison with the gold standard. For example, users were supposed to compute and enter the interval of years for an event mentioning *a woman in her 50s who travelled around* ... Some of them entered values corresponding to 1950 s.

Results reflect the fact that the way GATE presents annotations of event attributes often leads to inconsistent results. RDFaCE was only slightly better in this respect.

A part of the problem of event slots left empty although the annotated text contains information necessary for their filling (column "missing values") relates to pronominal references that were supposed to be linked to the referred entity. However, the difference between results of GATE and RDFaCE on one side and 4A on the other one shows that it is useful to visually distinguish system suggestions from user validated data and that 4A's way of confirming suggestions leads to more consistent data.

Finally, the average time needed to annotate an event was higher when our testers used RDFaCE than with the other two tools. This can be explained by a rather austere user interface of the tool with a limited way to easily correct previous mistakes.

4.4   Effects of Semantic Filtering

Previous results showed that even though an automatic annotation process cannot identify complex relations, it is beneficial to pre-annotate entities and basic relations and let users focus on high-level annotation tasks joining the prepared components and validating their linking to a knowledge base at the same time [5]. However, it has not been fully clear to what extent an indication of preferred types of attributes contributes to consistency of relation annotation and whether it improves annotator's comfort.

A set of 20 excerpts from documents on visual artworks (paintings, sculptures) and artistic influences were prepared for these experiments. Each of the texts mentioned several artworks, their authors and circumstances of their creation (dates, places, portrayed persons, etc.). There were also references to other artworks that inspired or influenced artists. The following paragraph shows a part of such a text:

> *Le déjeuner sur l'herbe is a large oil on canvas painting by Édouard Manet created in 1862 and 1863. Manet's composition reveals his study of the old masters, as the disposition of the main figures is derived from Marcantonio Raimondi's engraving of the Judgement of Paris (c. 1515).*

A group of 14 users identified mentions of artwork attributes in the texts first (mainly authors and creation dates). Then, they annotated influence relations between the artworks. Figure 2 demonstrates results of the process. Two configurations of the system were prepared. One highlighted potential semantic template values corresponding to the type of attribute being filled; the other one switched the semantic filtering function off. To exclude influences of the order of annotation, texts were presented to users in random order. Yet, the selection procedure guaranteed that each text will be annotated by 7 users with the semantic filtering function switched off and 7 with the function switched on.

Fig. 2. Artwork attributes and influence relations.

Table 2 compares annotation results obtained with the two settings. The 4A's semantic filtering switched on led to a higher quality of results. Relative decreases of the two types of errors exceed 25%. The annotation was also faster by 15%. Questionnaires that the annotators had filled immediately after the experiment also revealed that 11 out of 14 users had seen the semantic filtering as a feature significantly improving their experience, the other 3 agreed that it had helped them "moderately".

Table 2. Contribution of semantic filtering.

Semantic filtering	Incorrect values	Missing values	Time per relation
Switched off	6.9%	5.7%	41.4 s
Switched on	5.1%	4.2%	35.1 s

4.5   Optimizing Displayed Information

The last set of reported experiments explored the impact of varying amount of information presented to the user in an initial annotation view and the way users get additional information. It also asked whether users benefit from knowledge of potential alternative annotations and confidence levels of provided suggestions.

The experiments focused on complex entity disambiguation tasks. As mentioned above, the data extracted from the CommonCrawl corpus was searched for links that correspond to ambiguous names of people and places in the Wikipedia. A collection of 186 excerpts used in the tests was manually verified by one of the authors. The way it had been prepared guaranteed that a random guess would lead to a 50% error (or more, in the case of entities with more than 2 alternatives).

We primarily compared three settings of disambiguation views, differing by entity attributes shown, and looked at their impact on speed and accuracy of the disambiguation. Users were instructed to annotate just an entity in question (highlighted in each excerpt) and choose always one of provided suggestions. Users did not skip any disambiguation task so that we could compare just the speed and accuracy of results.

The first setting showed users an extensive list of attributes and values for suggestions with highest confidence values. Displayed attributes involved entity type, full name, description (corresponding to the first paragraph from Wikipedia or Freebase), visual representation (the first image from Wikipedia, if available), and URL. If necessary, users could follow the URL link, consult the relevant Wikipedia page and come to a decision based on the full information contained there.

The second setting corresponded to the limited view some tools offer for the disambiguation task. It displayed only entity types and URLs and users were supposed to either decide based on the URL alone, or open the Wikipedia page if they felt it is necessary for the disambiguation. Note that Wikipedia URLs can help disambiguation with words in parentheses used for articles discussing entities with the same name as a primary (more famous) entity covered by the Wikipedia.

The third setting took advantage of a special disambiguation attribute that is dynamically computed from descriptions of available alternatives. It combines disambiguation words from the Wikipedia URL and a selected part of the entity description. It is generated by a function which can be easily adapted to data sources differing from Wikipedia or Freebase. The disambiguation attribute was shown together with a suggested entity type and a URL so that users could again click to get more information.

While the sequence of testing cases (40 for each setting) was fixed, each of 6 testers had a different order of the 3 settings (similarly to the ordering of tools in the first set of experiments). Each user had 30 min for each setting. Table 3 compares times and error rates and shows how many times users clicked on the URL link to read further information.

Table 3. Results of experiments comparing three settings of the disambiguation view.

Setting	Average time	Error rate	URL clicked
Detailed information	33.92 s	6.2%	1.3%
Only type and URL	37.26 s	27.9%	41.7%
Disambiguation attr.	32.98 s	2.1%	1.5%

Though there were differences among individual testers, the overall figure (the best and the worst performing setting in terms of the average time and the error rate) remained the same for all testers. The number of cases in which individual users consulted Wikipedia pages was always the highest for the second setting but users differed in the level they believed that seeing just a URL is enough to decide (which then resulted in an increased number of errors).

The relatively high number of errors is also due to the complexity of the disambiguation task. This was one of the feedback answers provided by users after all 3 sessions in a questionnaire form. Although users tried to make as few mistakes as possible, 20+ minute sessions were felt demanding and users (not knowing how many errors they had made yet) pointed out that they could be faster if the focus would be on the speed rather than on the quality. Being confronted with the number of errors in their results, they realized the trade-off between the time and the quality and proposed context-sensitive features that would help them in particular disambiguation cases (images in the case of ambiguity between a ship name and a person, dates of deaths in the case of two persons living in different centuries, etc.).

The fact that users did not originally realize the complexity of the disambiguation task also probably explains the surprisingly high error in the case of presenting full information immediately (the first setting). Too much information that does not highlight key differences between alternatives seems to lead to a less focused work. Our future research will explore whether this can be changed when users are more experienced. On the other hand, the average time per decision and the connected low number of cases users had to consult Wikipedia pages correspond to the fact that users often skimmed full texts and images and felt they have enough information for their decisions.

The setting showing just the type and the URL proved to be the most diverse among users. Some of them opened more than 2/3 of all links and read the information on the Wikipedia page, others decided much faster but also made more errors. Although the latter could be prohibited by a penalization of errors, the second setting is clearly the worst for the task at hand. The tools that offer only this information in the disambiguation context could improve significantly by considering more informative views.

A clear winner of this part is the setting with the disambiguation attribute and the option to click on the provided URL to find details. Users made less mistakes than in other settings and the average time was the lowest. They needed to consult Wikipedia rarely. Five out of six users also indicated in the questionnaire that this setting was the most comfortable one in their eyes.

As opposed to the simplified situation prepared for the above-mentioned tested settings, the data for the next reported experiment corresponds to more realistic conditions when a name can belong to an entity that is not covered by a background knowledge base so that neither of the provided suggestions is correct. The focus on highly ambiguous names that have many alternative meanings in Wikipedia also prevents the simple selection strategy applicable in the previous settings. Users could not benefit from excluding the wrong alternative and thoughtlessly confirming the other one.

An experiment reported in [5] compared two settings of the disambiguation interface – one directly listing known entities sharing a name appearing in the text and the other one showing only the most probable candidate entity and letting users expand more alternatives by a click. The former showed to bring higher accuracy. That is why extended experiments covered by this paper present all alternatives to users and study how annotators perform in this case.

The number of entities sharing a given name was high (between 10 and 30) in selected texts. This corresponds to the situation when an automatic disambiguation engine has only limited information to decide so that confidence values for alternatives are small. The position of the correct choice varied in the data – 16 out of 20 texts included it in the list (on different positions) while in 4 remaining cases (20%), none of the alternatives corresponded to the entity actually mentioned in the text.

Two presentation forms of the list of alternatives were compared. The brief one listed only disambiguation contexts and enabled expanding a full list of entity attributes from the knowledge base (including description, visual representation, etc.) by a click. Figure 3 shows such a case. The expanded form listed all entity attributes directly – users had to browse through longer listings but they could easily match keywords from the text to the full entity descriptions instead.

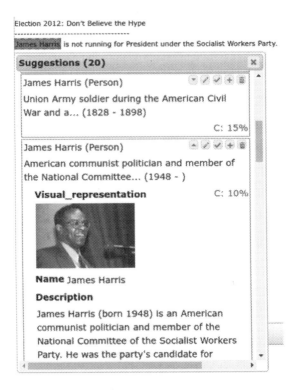

Fig. 3. An example of a full expansion of entity attributes for an alternative annotation.

Each of the 20 disambiguation tasks was solved by 7 annotators using the brief view of alternatives and other 7 dealing with the expanded one. Tasks for each user were grouped by the presentation form. One half of the users started with the expanded form; the other half with the brief one.

To cope with varying numbers of alternatives as well as changing positions of the correct ones, measured total times were always divided by the rank of the correct choice. This corresponds to the most frequently observed scenario (in 86% of the cases) in which users decided immediately when they had read an entity description matching the annotated text. Total numbers of alternatives were considered for the 4 cases of entities not covered by the knowledge base.

Results of the experiment are summarized in Table 4. The brief view enabled users to faster scan alternatives and to select the one of their choice. This was also confirmed by post-experiment questionnaires. All but one users preferred the brief view. The relatively high error rate corresponds to the complexity of the disambiguation task. As also suggested by questionnaire answers, the accuracy would probably increase if users do not stop at the alternative that seems to fit enough and consider all others in the list. However, this could slow down the disambiguation task more than two times according to the collected data. The potential decrease of performance seems to be too high to be acceptable.

Table 4. Disambiguation with brief and expanded views on alternative entity candidates.

Setting	Time per alternative	Error rate	Clicked to expand/collapse
Brief	11.2 s	12.1%	1.1%
Expanded	15.1 s	11.4%	5.3%

The last column of Table 4 characterizes interaction patterns observed when users worked with the list of alternatives. Only 1.1% candidate entity records in the brief mode were clicked to show the full description and other attributes. The brief disambiguation attribute has been mostly seen as comprehensive. On the other hand, users dealing with the expanded view clicked the collapse button in 5.3% of cases. The analysis of questionnaire answers then showed that this was mainly used to mark irrelevant choices. Our future research will look at these interaction patterns more closely.

5   Conclusions and Future Directions

Results of all three sets of experiments presented in this paper confirm a general finding – appropriate tool support for computer-assisted semantic annotation of text can bring significant advantages to the whole process – make it more consistent, faster and less demanding for users. If one considers a potential economic value of the manual preparation of annotation data (for example, to train advanced machine learning models for a particular complex task), it becomes critical to apply a tool with an optimal set of features for the particular annotation problem.

The empirical study focused on interaction components of semantic annotation systems suitable for complex relation annotation with highly ambiguous entities. Flexibility of the 4A system allowed us to switch on/off particular user interface components. It enabled evaluating relative contributions of specific features and showing those that bring clear benefits.

The 4A's semantic filter highlighting entities of expected types showed to be highly appreciated by users. It also led to a higher precision and slightly faster selection of attribute values. The experiments combining knowledge-based relation extraction (annotation of artwork authorship and creation dates followed by determining of artistic influences among the artworks mentioned in the text) also proved that clear distinction between system-generated annotation suggestions and user-confirmed annotations helps keep the result clear and prepare better training data for machine learning approaches.

Experiments comparing various settings of 4A's entity disambiguation interface proved that it is beneficial to pay a special attention to the amount of information presented to users in the case of entity name ambiguity and the way alternatives are presented. A brief context-dependent disambiguation text supplemented by the link to a Wikipedia page or another resource providing more details helped users to make fast and accurate decisions on the entity links.

Our future work will extend the reported results towards other kinds of complex annotation tasks including data preparation for aspect-oriented sentiment analysis and annotation of textual contexts suggesting emotional states of authors. We will also support newly available entity recognition tools and frameworks, such as WAT [13] or Gerbil [15], that will be employed as back-end pre-annotation components.

Acknowledgements. This work was supported by the H2020 project MixedEmotions, grant agreement No. 644632, and by The Ministry of Education, Youth and Sports of the Czech Republic from the National Programme of Sustainability (NPU II); project IT4Innovations Excellence in Science – LQ1602.

References

1. Bontcheva, K., Cunningham, H., Roberts, I., Roberts, A., Tablan, V., Aswani, N., Gorrell, G.: GATE Teamware: a web-based, collaborative text annotation framework. Lang. Resour. Eval. **47**(4), 1007–1029 (2013). https://doi.org/10.1007/s10579-013-9215-6
2. Bontcheva, K., Roberts, I., Derczynski, L., Rout, D.: The GATE crowdsourcing plugin: crowdsourcing annotated corpora made easy. In: Proceedings of Demonstrations at the 14th Conference of the European Chapter of the Association for Computational Linguistics (EACL), pp. 97–100. Association for Computational Linguistics (2014)
3. Ciccarese, P., Ocana, M., Clark, T.: Open semantic annotation of scientific publications using DOMEO. J. Biomed. Semant. 3(Suppl. 1) (2012). http://www.jbiomedsem.com/content/3/S1/S1
4. Cunningham, H., Maynard, D., Bontcheva, K., Tablan, V., Aswani, N., Roberts, I., Gorrell, G., Funk, A., Roberts, A., Damljanovic, D., Heitz, T., Greenwood, M.A., Saggion, H., Petrak, J., Li, Y., Peters, W.: Text Processing with GATE (Version 6). GATE (2011). http://tinyurl.com/gatebook
5. Dytrych, J., Smrz, P.: Interaction patterns in computer-assisted semantic annotation of text - an empirical evaluation. In: Proceedings of the 8th International Conference on Agents and Artificial Intelligence. SCITEPRESS, Rome, Italy (2016)
6. Grassi, M., Morbidoni, C., Nucci, M., Fonda, S., Donato, F.D.: Pundit: creating, exploring and consuming semantic annotations. In: Proceedings of the 3rd International Workshop on Semantic Digital Archives, Valletta, Malta (2013)
7. Handschuh, S., Staab, S., Ciravegna, F.: S-CREAM — semi-automatic CREAtion of metadata. In: Gómez-Pérez, A., Benjamins, V.R. (eds.) EKAW 2002. LNCS (LNAI), vol. 2473, pp. 358–372. Springer, Heidelberg (2002). https://doi.org/10.1007/3-540-45810-7_32
8. Heese, R., Luczak-Rsch, M., Paschke, A., Oldakowski, R., Streibel, O.: One click annotation. In: Proceedings of the 6th Workshop on Scripting and Development for the Semantic Web, collocated with ESWC. Ruzica Piskac, Redaktion Sun SITE, Informatik V, RWTH Aachen, Ahornstr. 55, 52056 Aachen, Germany (2010)
9. Khalili, A., Auer, S., Hladky, D.: The RDFa Content Editor - From WYSIWYG to WYSIWYM. In: Proceedings of COMPSAC 2012 - Trustworthy Software Systems for the Digital Society (2012). http://svn.aksw.org/papers/2012/COMPSAC_RDFaCE/public.pdf

10. Maynard, D.: Benchmarking textual annotation tools for the semantic web. In: 6th International Conference on Language Resources and Evaluation. European Language Resources Association (ELRA), Marrakech, Morocco (2008). https:// gate.ac.uk/sale/lrec2008/benchmarking.pdf

11. Maynard, D., Dasiopoulou, S., Costache, S., Eckert, K., Stuckenschmidt, H., Dzbor, M., Handschuh, S.: Knowledge web project: Deliverable D1.2.2.1.3 - Benchmarking of annotation tools (2007). http://knowledgeweb.semanticweb.org/ semanticportal/deliverables/D1.2.2.1.3.pdf

12. Moro, A., Navigli, R.: SemEval-2015 Task 13: Multilingual all-words sense disambiguation and entity linking. In: Proceedings of the 9th International Workshop on Semantic Evaluation, Denver, Colorado, pp. 288–297, June 2015. http://www. aclweb.org/anthology/S15-2049

13. Piccinno, F., Ferragina, P.: From TagME to WAT: a new entity annotator. In: Proceedings of the First International Workshop on Entity Recognition & Disambiguation, pp. 55–62. ACM (2014)

14. Reeve, L., Han, H.: Survey of semantic annotation platforms. In: Proceedings of the 2005 ACM Symposium on Applied Computing, SAC 2005, pp. 1634–1638. ACM, New York (2005). https://doi.org/10.1145/1066677.1067049

15. Röder, M., Usbeck, R., Ngomo, A.-C.N.: Developing a sustainable platform for entity annotation benchmarks. In: Gandon, F., Guéret, C., Villata, S., Breslin, J., Faron-Zucker, C., Zimmermann, A. (eds.) ESWC 2015. LNCS, vol. 9341, pp. 190–196. Springer, Cham (2015). https://doi.org/10.1007/978-3-319-25639-9_36. http://svn.aksw.org/papers/2015/ESWC_GERBIL_semdev/public.pdf

16. Smrz, P., Dytrych, J.: Towards new scholarly communication: A case study of the 4A framework. In: SePublica. CEUR Workshop Proceedings, vol. 721. Ruzica Piskac, Redaktion Sun SITE, Informatik V, RWTH Aachen, Ahornstr. 55, 52056 Aachen, Germany (2011)

17. Smrz, P., Dytrych, J.: Advanced features of collaborative semantic annotators - the 4A system. In: Proceedings of the 28th International FLAIRS Conference. AAAI Press, Hollywood, Florida, USA (2015)

18. Stenetorp, P., Pyysalo, S., Topić, G., Ohta, T., Ananiadou, S., Tsujii, J.: BRAT: a web-based tool for NLP-assisted text annotation. In: Proceedings of the Demonstrations at the 13th Conference of the European Chapter of the Association for Computational Linguistics, EACL 12, pp. 102–107 (2012). Association for Computational Linguistics, Stroudsburg, PA, USA. http://dl.acm.org/citation.cfm? id=2380921.2380942

19. Surdeanu, M., Heng, J.: Overview of the English slot filling track at the TAC2014 knowledge base population evaluation. In: Proceedings of the TAC-KBP 2014 Workshop (2014)

20. Wang, A., Hoang, C., Kan, M.Y.: Perspectives on crowdsourcing annotations for natural language processing. Lang. Resour. Eval. 47(1), 9–31 (2013). https://doi. org/10.1007/s10579-012-9176-1

21. Yee, K.P.: Critlink: Advanced hyperlinks enable public annotation on the web (2002). http://zesty.ca/pubs/cscw-2002-crit.pdf, http://zesty.ca/pubs/cscw-2002-crit.pdf

Author Index

Alves, Victor   48

Badie, Farshad   1
Bergenti, Federico   157
Bleiweiss, Avi   82
Boyarski, Eli   116

Dytrych, Jaroslav   205

Faria, Ricardo   48
Felner, Ariel   116
Ferraz, Filipa   48

Geihs, Kurt   22

Iotti, Eleonora   157

Jakob, Stefan   22

Komenda, Antonín   137
Král, Pavel   102

Monica, Stefania   157

Neves, João   48
Neves, José   48

Ohsuga, Akihiko   62
Opfer, Stephan   22
Orihara, Ryohei   62

Poggi, Agostino   157

Racharak, Teeradaj   179
Rajtmajer, Václav   102

Sato, Minato   62
Sei, Yuichi   62
Smrz, Pavel   205
Štolba, Michal   137
Surynek, Pavel   116
Švancara, Jiří   116

Tahara, Yasuyuki   62
Tojo, Satoshi   179
Tožička, Jan   137

Vicente, Henrique   48

Printed in the United States
By Bookmasters